Synthesis Lectures on Advances in Automotive Technology

Series Editor

Amir Khajepour, University of Waterloo, Waterloo, ON, Canada

Editorial Board

Akram A. Almohammedi, South Ural State University, Chelyabinsk, Russia
Albert P. C. Chan, Hong Kong Polytechnic University, Kowloon, Hong Kong
Anand Nayyar, Duy Tan University, Da Nang, Vietnam
Andrea Appolloni, University of Rome Tor Vergata, Rome, Italy
Baochang Zhang, Beihang University, Beijing, China
Behnam Mohammadi-Ivatloo, University of Tabriz, Tabriz, Iran
Bingbing Gao, Northwestern Polytechnical University, Xi'an, China
Chi-Cheng Chang, National Taiwan Normal University, Taipei, Taiwan
Chi-Hua Chen, Fuzhou University, Fuzhou, China
Christina Zong-Hao Ma, Hong Kong Polytechnic University, Kowloon, Hong Kong
Chuan-Pei Lee, University of Taipei, Taipei, Taiwan
Chuanyin Xiong, Shaanxi University of Science and Technology, Xianyang, China
David Bassir, Paris-Saclay University, Paris, France
Debashish Bhowmik, Tripura Institute of Technology, Narsingarh, India
Dewu Shu, Shanghai Jiaotong University, Shanghai, China
Dragan Pamucar, University of Defence, Belgrade, Serbia
Ellips Masehian, California State Polytechnic University, Pomona, USA
Francesco Caracciolo, University of Naples Federico II, Naples, Italy
Francisco Perlas Dumanig, University of Hawai'i, at Hilo, USA
Gabriella Pasi, University of Milano-Bicocca, Milan, Italy
Gilbert Nartea, University of Canterbury, Christchurch, New Zealand
Gordon Huang, Beijing Normal University, Beijing, China
Gu He, Sichuan University, Chengdu, China
Guiyin Li, Guilin University of Electronic Technology, Guilin, China
Guoqing Chen, Tsinghua University School of Economics and Management, Beijing, China
Hexiu Xu, Air Force Engineering University, Xi'an, China
Huang Weimin, Nanyang Technological University, Jurong West, Singapore
Humberto Bustince, Public University of Navarra, Pamplona, Spain
Hwai Chyuan Ong, University of Technology Sydney, Sydney, Australia
Jessica M. Black, Boston College, Chestnut Hill, USA
Jia Li, Hunan University, Changsha, China
Jing Liu, Hebei University of Technology, Tianjin, China
Jiuqing Cheng, University of Northern Iowa, Cedar Falls, USA
Jun Yang, Lanzhou Institute of Chemical Physics, Chinese Academy of Sciences, Lanzhou, China
Juntao Tang, Central South University, Changsha, China
Junwei Wu, Harbin Institute of Technology, Shenzhen, China
Kan Li, Beijing Institute of Technology, Beijing, China
Khalil Khan, University of Azad Jammu and Kashmir, Muzfarabbad, Pakistan
Long Wang, University of Science and Technology Beijing, Beijing, China
Mathias Urban, Dublin City University, Dublin, Ireland

Mohamed Arezki Mellal, University of Boumerdès, Boumerdès, Algeria
Mohammad T. Khasawneh, State University of New York at Binghamton, New York, USA
Nancy Scheper-Hughes, University of California Berkeley, Berkeley, USA
Periklis Gogas, Democritus University of Thrace, Komotini, Greece
Peter Sachsenmeier, Hankou University, Wuhan, China
Philippe Fournier-Viger, Harbin Institute of Technology, Shenzhen, China
Qijun Sun, Beijing Institute of Nanoenergy and Nanosystems, Chinese Academy of Sciences, Beijing, China
Qingyong Li, Beijing Jiaotong University, Beijing, China
Quang Ngoc Quang, Waseda University, Tokyo, Japan
Quanxin Zhu, Hunan Normal University, Changsha, China
Radhi Al-Mabuk, University of Northern Iowa, Cedar Falls, USA
Radko Mesiar, Slovak University of Technology in Bratislava, Bratislava, Slovakia
Ramadas Narayanan, Central Queensland University, Bundaberg, Australia
Rongyi Chen, Guangdong Medical University, Zhanjiang, China
Runliang Dou, Tianjin University, Tianjin, China
Sayantan Mazumdar, Nankai University, Tianjin, China
Sharif Ullah, Kitami Institute of Technology, Kitami, Japan
Shengxiong Xiao, Shanghai Normal University, Shanghai, China
Siqian Gong, Beijing Jiaotong University, Beijing, China
Stephen A. Butterfield, University of Maine, Orono, USA
Stephen Webb, Glasgow Caledonian University, Glasgow, UK
Steven Guan, Xi'an Jiaotong-Liverpool University, Suzhou, China
Thanh Ngo, Massey University, Palmerston North, New Zealand
Valentina Emilia Balas, Aurel Vlaicu University of Arad, Arad, Romania
Wadim Strielkowski, Centre for Energy Studies, Prague Business School, Prague, Czech Republic
Wanshu Ma, Northwestern University, Chicago, USA
Wei Zhang, Beijing University of Technology, Beijing, China
Wei Zheng, Northwestern Polytechnical University, Xi'an, China
Wei-Chiang Hong, Jiangsu Normal University, Xuzhou, China
Xianfei Ding, First Affiliated Hospital of Zhengzhou University, Zhengzhou, China
Xuesong Xu, Hunan University of Technology and Business, Changsha, China
Xun Liang, Renmin University of China, Beijing, China
Xun Luo, Tianjin University of Technology, Tianjin, China
Yongchao Jiang, Zhengzhou University, Zhengzhou, China
Zaicheng Sun, Beijing University of Technology, Beijing, China
Zheng Zheng, Beihang University, Beijing, China
Zhiyu Xi, Beihang University, Beijing, China
Zhuoqi Ding, Shenzhen Technology University, Shenzhen, China
Zuhua Wang, Guiyang College of Traditional Chinese Medicine, Guiyang, China

This series covers the significant advances in new manufacturing techniques, low-cost sensors, high processing power, and ubiquitous real-time access to information that mean vehicles are rapidly changing and growing in complexity. These new technologies (including the inevitable evolution toward autonomous vehicles) will ultimately deliver substantial benefits to drivers, passengers and the environment. These publications cover the cutting edge of advanced automotive technologies.

Yukun Lu · Chen Sun · Amir Khajepour

Active and Semi-active Suspension Systems

Modeling, Control, and Fault Diagnosis

Yukun Lu
Department of Mechanical and Mechatronics
Engineering
University of Waterloo
Waterloo, ON, Canada

Chen Sun
Department of Mechanical and Mechatronics
Engineering
University of Waterloo
Waterloo, ON, Canada

Amir Khajepour
Department of Mechanical and Mechatronics
Engineering
University of Waterloo
Waterloo, ON, Canada

ISSN 2576-8107 ISSN 2576-8131 (electronic)
Synthesis Lectures on Advances in Automotive Technology
ISBN 978-3-031-73923-1 ISBN 978-3-031-73924-8 (eBook)
https://doi.org/10.1007/978-3-031-73924-8

© The Editor(s) (if applicable) and The Author(s), under exclusive license to Springer
Nature Switzerland AG 2025

This work is subject to copyright. All rights are solely and exclusively licensed by the Publisher, whether the whole or part of the material is concerned, specifically the rights of translation, reprinting, reuse of illustrations, recitation, broadcasting, reproduction on microfilms or in any other physical way, and transmission or information storage and retrieval, electronic adaptation, computer software, or by similar or dissimilar methodology now known or hereafter developed.
The use of general descriptive names, registered names, trademarks, service marks, etc. in this publication does not imply, even in the absence of a specific statement, that such names are exempt from the relevant protective laws and regulations and therefore free for general use.
The publisher, the authors and the editors are safe to assume that the advice and information in this book are believed to be true and accurate at the date of publication. Neither the publisher nor the authors or the editors give a warranty, expressed or implied, with respect to the material contained herein or for any errors or omissions that may have been made. The publisher remains neutral with regard to jurisdictional claims in published maps and institutional affiliations.

This Springer imprint is published by the registered company Springer Nature Switzerland AG
The registered company address is: Gewerbestrasse 11, 6330 Cham, Switzerland

If disposing of this product, please recycle the paper.

Preface

This book explores the essential aspects of active and semi-active suspension systems, providing a clear and accessible approach to their understanding and design. Specifically intended for undergraduate students, it is also suitable for anyone interested in learning about the foundations and design of suspension control and diagnostic systems. The relevant practical implantations for sedan, SUV, and truck systems are given in examples and simulated in MATLAB and CarSim software. Utilizing a step-by-step method enriched with pictures, graphs, codes, and examples, the reader is guided through complex concepts with ease.

Chapter 1 introduces semi-active and active suspension systems, discussing their advantages and presenting a brief overview of some mass-produced adaptive suspension systems available in the market. Chapter 2 delves into three widely used vehicle suspension modelling methods. Chapter 3 gives a review of existing suspension control techniques. Chapter 4 offers an in-depth examination of several popular control algorithms used in adaptive suspension systems. The book ends with Chapter 5, which focuses on the diagnosis and prognosis of suspension systems.

Waterloo, Canada
Yukun Lu
Chen Sun
Amir Khajepour

Acknowledgment This book would not have been possible without the help of many people. We are particularly grateful to Michael Duthie, Jeff Graansma, and Aaron Sherratt from the Mechatronic Vehicle Systems Lab at the University of Waterloo, for their valuable assistance. We are also thankful to Springer Nature for providing the publishing opportunity and for their consistent encouragement and support throughout this project.

Contents

1 **Introduction** 1
   1.1 Why Are Advanced Suspension Systems Needed? 1
   1.2 Active Suspensions 3
   1.3 Semi-Active Suspensions 4
   1.4 Common Mass-Produced Active and Semi-Active Suspensions 4

2 **Mathematical Modeling of Suspension Systems** 7
   2.1 Quarter-Car Model 7
   2.2 Half-Car Model 9
       2.2.1 Vertical and Pitch 9
       2.2.2 Vertical and Roll 10
   2.3 Full-Car Model 11

3 **Common Suspension Control Techniques** 15
   3.1 Model-Based Control 18
   3.2 Gain-Adaptive Approach 19

4 **Suspension Control Applications** 23
   4.1 Skyhook Control 23
   4.2 Linear Quadratic Regulator (LQR) 28
   4.3 Model Predictive Control (MPC) 33
   4.4 Integrated Skyhook-LQR 44
   4.5 Gain-Adaptive Algorithms 48

5 **Diagnosis and Prognosis of Suspension Systems** 53
   5.1 Common Component Failures in Suspension Systems 54
   5.2 Suspension Fault Detection and Identification 54
       5.2.1 Fault Detection with Non-Recursive Parameter Estimation 54
       5.2.2 Fault Detection with Recursive Parameter Estimation 57

5.3	Suspension Fault Diagnosis and Prognosis		59
	5.3.1 Fault Diagnosis Procedure		59
	5.3.2 Outlooks on Fault Prognosis		61

References 63

Introduction

Vehicle suspension systems play a critical role in ensuring passenger comfort, vehicle stability, and overall handling characteristics. By effectively managing the interaction between the vehicle and the road surface, suspension systems contribute to a smoother ride, improved traction, and enhanced safety. With advancements in automotive technology and the growing demand for superior ride quality and performance, vehicle suspension control has emerged as a key area of research and development. The primary objective of vehicle suspension control is to optimize suspension behavior in response to changing road conditions, vehicle dynamics, and driver inputs. Traditional passive suspension systems provide fixed damping characteristics, which may not adequately address varying road surfaces or dynamic driving maneuvers. In contrast, modern active and semi-active suspension systems utilize advanced control algorithms, actuators, and sensors to continuously adjust suspension parameters in real time, thereby enhancing ride comfort, handling, and stability. As automotive technology advances, the importance of sophisticated suspension systems will only grow, making them a crucial component in the future landscape of transportation.

1.1 Why Are Advanced Suspension Systems Needed?

The primary objective of vehicle suspension control is to optimize suspension behavior in response to changing road conditions, vehicle dynamics, and driver inputs. Traditional passive suspension systems provide fixed damping characteristics, which may not adequately address varying road surfaces or dynamic driving maneuvers. In contrast, advanced active and semi-active suspension systems utilize advanced control algorithms, actuators, and sensors to continuously adjust suspension parameters in real time. Active suspension systems go a step further by actively controlling the suspension components using hydraulic or electromagnetic actuators. These systems can significantly reduce body roll, pitch, and bounce, offering an unparalleled driving experience by maintaining optimal tire contact with the road surface under all conditions. Semi-active suspension systems, while less complex, still offer considerable improvements over passive systems by adjusting damping levels to suit different driving scenarios.

Thereby, active and semi-active suspension systems can address several key challenges and requirements in modern automotive design and performance:

- Enhanced Ride Comfort: Advanced suspension systems can provide superior ride comfort by effectively isolating passengers from road disturbances, bumps, and vibrations. By incorporating adaptive damping, active control, and predictive algorithms, these systems can adjust suspension parameters in real-time to minimize vehicle body motions and optimize ride quality.
- Improved Handling and Stability: Advanced suspension systems can enhance vehicle handling and stability by optimizing tire contact with the road surface and controlling body roll and pitch motions. Active suspension systems can adjust damping rates, stiffness, and ride height to improve cornering performance, minimize body roll during turns, and maintain stability under varying driving conditions.
- Increased Safety: Advanced suspension systems can contribute to vehicle safety by improving stability, handling, and control in emergency situations and adverse driving conditions. By minimizing body roll, reducing vehicle pitch and dive, and maintaining tire contact with the road surface, these systems help prevent loss of control, skidding, and rollover accidents, thereby enhancing occupant safety.
- Adaptability to Various Road Conditions: Advanced suspension systems can adapt to a wide range of road conditions, surfaces, and driving environments, ensuring optimal performance and comfort across diverse operating scenarios. By adjusting suspension parameters based on real-time inputs from sensors and vehicle dynamics, these systems can provide consistent and predictable behavior on smooth highways, rough roads, and off-road terrain.
- Integration with Advanced Driver Assistance Systems (ADAS): Advanced suspension systems can seamlessly integrate with ADAS technologies, such as lane-keeping assist, adaptive cruise control, and collision avoidance systems, to enhance overall vehicle safety and automation capabilities. By coordinating suspension adjustments with other vehicle systems, these systems can improve vehicle stability, control, and responsiveness in conjunction with driver assistance features.

Moreover, the evolution of vehicle suspension systems is closely linked with advancements in other automotive technologies, such as autonomous driving and electric vehicles (EVs). Autonomous vehicles require suspension systems that can seamlessly adapt to a variety of conditions without human intervention, ensuring safety and comfort for passengers. For EVs, the challenge lies in balancing the weight of the battery packs with the need for efficient and responsive suspension control. The integration of intelligent control systems and the use of artificial intelligence (AI) in suspension technology represent the forefront of current research. AI-driven systems can learn and predict optimal suspension settings based on a wide range of variables, including vehicle speed, load distribution, and road surface characteristics. These systems promise to further elevate vehicle performance and passenger comfort to new heights.

Future developments in suspension technology also hold potential benefits for specialized applications, such as off-road vehicles, commercial trucks, and high-performance sports cars. Off-road vehicles can benefit from suspension systems that provide superior articulation and durability, while commercial trucks can achieve better load management and reduced driver fatigue. High-performance sports cars, on the other hand, can push the limits of speed and agility with precision-tuned suspension systems that offer maximum grip and control.

1.2 Active Suspensions

Active suspension systems differ fundamentally from traditional or passive suspension systems. In a passive setup, energy is stored in a spring and dissipated through a shock absorber, which has fixed damping characteristics. This means the system cannot adapt to changing road conditions or driving maneuvers, leading to compromises in comfort, handling, and stability. While active suspension systems replace passive shock absorbers with high-speed actuators. These actuators, controlled by advanced algorithms and sensors, continuously adjust the suspension parameters in real-time. This dynamic adaptability allows the vehicle to maintain optimal ride quality and handling, regardless of road conditions or driving style. Specifically, this system can independently control each wheel's suspension, providing the highest level of performance and adaptability. It is capable of managing body roll, pitch, and bounce over a wide range of frequencies (0 to 30 Hz).

The key components of a typical active suspension system include:

- Actuators: The core of an active suspension system, these devices can apply varying levels of force to the suspension components, allowing for real-time adjustments.
- Sensors: These monitor a range of parameters including wheel speed, body acceleration, and road conditions. The data collected is crucial for the system to make informed adjustments.
- Electric Control Unit: The brain of the system, which processes sensor data and determines the necessary adjustments. This unit employs sophisticated algorithms to balance comfort, handling, and stability.

- Power Supply: Active suspensions require a significant amount of power to operate the actuators, often supplied by the vehicle's electrical system or dedicated power sources in hybrid configurations.

While active suspension systems offer numerous advantages, they also come with challenges. The high cost of development and implementation, significant power requirements, and increased system complexity have historically limited their widespread adoption. However, ongoing advancements in materials science, control algorithms, and energy-efficient technologies are addressing these issues, making active suspensions more accessible and practical for a broader range of vehicles. As the automotive industry evolves towards greater levels of integration, automation, and efficiency, active suspension systems are poised to play a crucial role. They offer a pathway to achieving unprecedented levels of vehicle dynamics, providing drivers and passengers with unparalleled comfort, safety, and performance.

1.3 Semi-Active Suspensions

Semi-active suspension systems represent an intermediate technology that adjusts the damping characteristics of the shock absorbers in real-time in response to road conditions and vehicle dynamics. Unlike fully active suspension systems, which use high-speed actuators to control the suspension components independently, semi-active systems modulate the damper settings to adapt to the driving environment. This adaptability enhances ride comfort, handling, and stability without the high cost and power consumption associated with fully active systems.

- Magnetorheological (MR) Dampers: These dampers contain a fluid whose viscosity can be altered by applying a magnetic field. By changing the viscosity, the damping characteristics can be adjusted in real-time, allowing for rapid adaptation to road conditions.
- Continuous Damping Control (CDC) Dampers: These dampers feature a variable damping valve that can adjust the flow of hydraulic fluid within the damper. By modifying this flow, the damping force can be altered continuously and in real-time.

1.4 Common Mass-Produced Active and Semi-Active Suspensions

Active and semi-active suspension systems have revolutionized the automotive industry by providing enhanced ride comfort, handling, and safety. Here are some common mass-produced adaptive suspension systems:

1.4 Common Mass-Produced Active and Semi-Active Suspensions

- Audi Magnetic Ride[1]: Audi's Magnetic Ride system is a popular active suspension technology used in various models, including the Audi TT and Audi R8. This system uses magnetorheological (MR) dampers, allowing for rapid adaptation to road conditions. This system enhances both ride comfort and handling precision.
- Mercedes-Benz AIRMATIC[2]: Mercedes-Benz AIRMATIC suspension is a widely used air suspension system that combines air springs with adaptive dampers. This system can automatically adjust the ride height and damping characteristics based on driving conditions and vehicle load. The AIRMATIC system improves ride comfort, handling stability, and can raise or lower the vehicle for better aerodynamics or off-road capability.
- BMW Adaptive M Suspension[3]: BMW's Adaptive M Suspension is an active system used in performance-oriented models. This system features electronically controlled dampers that continuously adjust the damping force based on real-time inputs from various sensors. The Adaptive M Suspension enhances cornering performance, ride comfort, and overall driving dynamics.
- Cadillac Magnetic Ride Control[4]: Cadillac's Magnetic Ride Control is one of the fastest-reacting suspension systems in the industry. Similar to Audi's Magnetic Ride, this system uses MR dampers filled with magnetorheological fluid. By applying a magnetic field, the system can adjust the damping force up to 1,000 times per second. This technology is employed in several Cadillac models, including the Escalade and the CT5-V, providing a balance between sporty handling and comfortable ride quality.
- Tesla Smart Air Suspension[5]: Tesla's Smart Air Suspension is featured in models like the Model S and Model X. This system uses air springs and electronically controlled dampers to automatically adjust the ride height and damping based on driving conditions and terrain. The Smart Air Suspension enhances ride comfort, efficiency, and aerodynamics by lowering the vehicle at high speeds and raising it for rough roads or obstacles.
- Land Rover Terrain Response[6]: Land Rover's Terrain Response system is an advanced active suspension technology used in their off-road capable vehicles, such as the Range Rover and Discovery. This system allows the driver to select different modes based on the terrain, such as grass, gravel, snow, mud, ruts, sand, and rock crawl. The suspension system then adjusts the ride height, damping, and other parameters to optimize

---

[1] https://www.audi.ca/ca/web/en/search-terms/Audi-Magnetic-Ride.html.

[2] https://www.fcpeuro.com/blog/the-definitive-guide-to-the-mercedes-benz-airmatic-suspension-system.

[3] https://www.bmw-m.com/en/topics/magazine-article-pool/das-adaptive-M-fahrwerk.html.

[4] https://www.germaincadillacofeaston.com/cadillac-magnetic-ride-control/.

[5] https://www.tesla.com/ownersmanual/models/en_us/GUID-F1B6801A-8946-41AD-8CF9-7A963CDA38E4.html.

[6] https://www.landroverusa.com/our-story/terrain-response.html.

performance for the selected terrain, providing exceptional off-road capability and ride comfort.

These mass-produced adaptive suspension systems illustrate the significant advancements in automotive suspension technology. By incorporating real-time adaptability and advanced control mechanisms, these systems provide a superior driving experience that combines comfort, handling, and safety. As automotive technology continues to evolve, fully active suspensions are expected to become even more sophisticated and widespread, further enhancing vehicle performance and driver satisfaction.

Mathematical Modeling of Suspension Systems

Mathematical modeling of vehicle suspensions is essential in understanding and optimizing their performance characteristics. The development of mathematical models for vehicle suspensions involves capturing the dynamic interactions between suspension components, vehicle body, and road surface. One of the primary objectives of vehicle suspension mathematical modeling is to predict and understand the dynamic response of the suspension system to external inputs, such as road disturbances, vehicle maneuvers, and driver commands. By quantifying the relationship between suspension parameters, vehicle dynamics, and performance metrics such as ride comfort, handling, and stability, mathematical models provide valuable insights into the behavior of suspension systems across a wide range of operating conditions. Mathematical models of vehicle suspensions can take various forms, ranging from simple analytical models to complex multi-body dynamic simulations. Analytical models, such as single-degree-of-freedom or quarter-car models, offer simplified representations of suspension behavior and are commonly used for preliminary analysis and design. More sophisticated models, such as multi-body dynamic models or finite element models, provide detailed insights into the interactions between suspension components, vehicle structure, and road surface, enabling more comprehensive analysis and optimization.

2.1 Quarter-Car Model

A quarter-car model has 2 DOFs is introduced and modelled in this section. As shown in Fig. 2.1, it considers the vertical motions of both unsprung mass m_u and sprung mass m_s. The unsprung mass represents the wheel and its associated components, and the sprung

Fig. 2.1 Quarter-car models

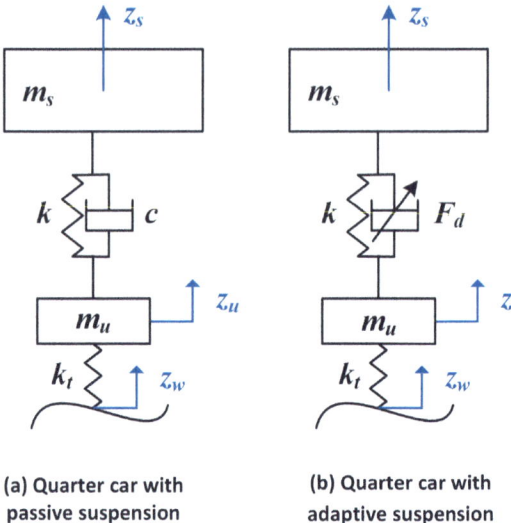

(a) Quarter car with passive suspension

(b) Quarter car with adaptive suspension

mass represents the corresponding vehicle's body mass to the wheel (approximately a quarter of the body mass). Sprung and unsprung mass' vertical motions are described by z_s and z_u, respectively. The stiffness and damping coefficient of the passive suspension in Fig. 2.1a are represented by k and c, respectively. The tire vertical stiffness is k_t and its damping is usually negligible with respect to the suspension damping. The damping characteristic is adjustable in the adaptive suspension system, which is usually modelled as a controllable damping force F_d, as shown in Fig. 2.1b.

Considering the passive suspension system in Fig. 2.1a, its differential equations of sprung and unsprung mass' vertical motions can be obtained using Newton's second law:

$$m_s \ddot{z}_s = c(\dot{z}_u - \dot{z}_s) + k(z_u - z_s)$$
$$m_u \ddot{z}_u = -c(\dot{z}_u - \dot{z}_s) - k(z_u - z_s) - k_t(z_u - z_w) \qquad (2.1)$$

The above equations can be written in the state-space form as follows:

$$\underbrace{\begin{bmatrix} \dot{z}_s \\ \ddot{z}_s \\ \dot{z}_u \\ \ddot{z}_u \end{bmatrix}}_{\dot{X}} = \underbrace{\begin{bmatrix} 0 & 1 & 0 & 0 \\ -\frac{k}{m_s} & -\frac{c}{m_s} & \frac{k}{m_s} & \frac{c}{m_s} \\ 0 & 0 & 0 & 1 \\ \frac{k}{m_u} & \frac{c}{m_u} & -\frac{k+k_t}{m_u} & -\frac{c}{m_u} \end{bmatrix}}_{A} \underbrace{\begin{bmatrix} z_s \\ \dot{z}_s \\ z_u \\ \dot{z}_u \end{bmatrix}}_{X} + \underbrace{\begin{bmatrix} 0 \\ 0 \\ 0 \\ \frac{k_t}{m_u} \end{bmatrix}}_{B} \underbrace{z_w}_{U} \qquad (2.2)$$

As for the adaptive suspension system in Fig. 2.1b, the damping term $c(\dot{z}_u - \dot{z}_s)$ can be represented by a force term F_d generated by the adaptive damper, which will be introduced in detail in the next section.

2.2 Half-Car Model

The vehicle quarter-car model is a fundamental tool in the study and design of vehicle suspension systems. By providing insights into the dynamic response of a single wheel and its suspension, engineers can optimize the suspension parameters to achieve a balance between ride comfort, road holding, and suspension travel, thereby improving the overall performance of the vehicle.

2.2 Half-Car Model

The vehicle half-car model is a more detailed representation of a vehicle's suspension system than the quarter-car model. It considers two DOFs of the vehicle body (vertical and pitch, or vertical and roll), providing a more comprehensive analysis of the vehicle's dynamic behavior.

2.2.1 Vertical and Pitch

A half-car model includes the vertical and pitch motions of the body is shown in Fig. 2.2. The differential equations of the sprung mass' vertical and pitch motions can be written as:

$$m_s \ddot{z}_s = k_1(z_{u1} - z_{s1}) + k_2(z_{u2} - z_{s2}) + F_1 + F_2$$
$$I_y \ddot{\theta}_s = -k_1 L_1(z_{u1} - z_{s1}) + k_2 L_2(z_{u2} - z_{s2}) - F_1 L_1 + F_2 L_2 - m_s a_x h \quad (2.3)$$

in which m_s represents the weight of the sprung mass; I_y is the sprung mass moment of inertia around the y-axis; L_1 and L_2 are the distances from the vehicle CG location to the front and rear axles, respectively; a_x is the vehicle longitudinal acceleration; h is the vehicle CG height; and F_i represents the damping forces as follows:

$$F_i = c_i(\dot{z}_{ui} - \dot{z}_{si}) \quad (2.4)$$

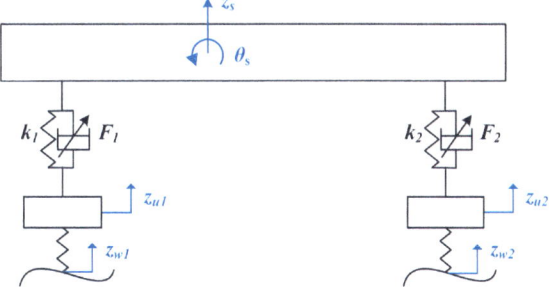

Fig. 2.2 Half-car model considers vertical and pitch motions

Besides, the suspension relative displacements can be simplified according to the small-angle simplifications:

$$z_{s1} = z_s - L_1\theta_s$$
$$z_{s2} = z_s + L_2\theta_s \qquad (2.5)$$

2.2.2 Vertical and Roll

A half-car model includes the vertical and roll motions of the body is shown in Fig. 2.3. The differential equations of the sprung mass' vertical and roll motions can be written as:

$$m_s\ddot{z}_s = k_1(z_{u1} - z_{s1}) + k_3(z_{u3} - z_{s3}) + F_1 + F_3$$
$$I_x\ddot{\varphi}_s = k_1l(z_{u1} - z_{s1}) - k_3l(z_{u3} - z_{s3}) + F_1l - F_3l - \tau_{bar} + m_s a_y h \qquad (2.6)$$

in which I_x is the sprung mass moment of inertia around the x-axis; a_y is the vehicle lateral acceleration; l is the half track-width; and F_i represents the damping forces as follows:

$$F_i = c_i(\dot{z}_{ui} - \dot{z}_{si}) \qquad (2.7)$$

The torque generated by the anti-roll bar can be calculated by

$$\tau_{bar} = k_{bar}\left(\varphi_s - \frac{1}{2l}z_{u1} + \frac{1}{2l}z_{u3}\right) \qquad (2.8)$$

Besides, the suspension relative displacements can be simplified according to the small-angle simplifications:

$$z_{s1} = z_s + l\varphi_s$$
$$z_{s3} = z_s - l\varphi_s \qquad (2.9)$$

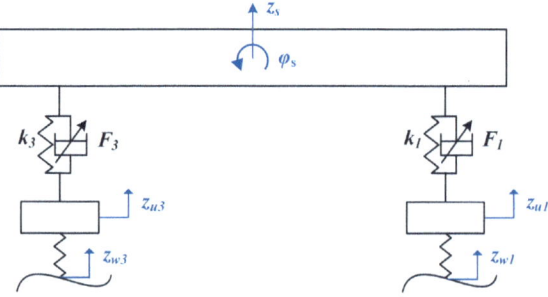

Fig. 2.3 Half-car model considers vertical and roll motions

2.3 Full-Car Model

The vehicle half-car model provides a detailed analysis of the vehicle's suspension dynamics, which is essential for designing and optimizing suspension systems to achieve a balance between ride comfort, handling, and stability.

2.3 Full-Car Model

The vehicle full-car model is a comprehensive mathematical representation of the suspension system. Unlike the quarter-car or half-car models, the full-car model considers the dynamic interactions of all four wheels as shown in Fig. 2.4a, providing a detailed analysis of the vehicle's vertical, roll, and pitch motions. Figure 2.4b shows a simplified full-car model with 7 DOFs. The right-handed Cartesian coordinate system is used to describe the vehicle system. z_s is the sprung mass vertical displacement which is defined as deviation from the nominal position. φ_s and θ_s are the roll and pitch angle of the vehicle body. z_{si} represents the vertical displacement of the upper installation point of the suspension system, and z_{ui} represents the vertical displacement of the unsprung mass. The four corners of the suspension system are distinguished by i ($i = 1, 2, 3, 4$).

The dynamics of sprung mass can be represented by the following equations:

$$m_s \ddot{z}_s = F_{s1} + F_{s2} + F_{s3} + F_{s4} + F_1 + F_2 + F_3 + F_4$$
$$I_x \ddot{\varphi}_s = (F_{s1} + F_1 + F_{s2} + F_2)l - (F_{s3} + F_3 + F_{s4} + F_4)l - \tau_{bar1} - \tau_{bar2} + m_s a_y h$$
$$I_y \ddot{\theta}_s = -(F_{s1} + F_1 + F_{s3} + F_3)L_1 + (F_{s2} + F_2 + F_{s4} + F_4)L_2 - m_s a_x h \quad (2.10)$$

in which I_x and I_y are the sprung mass moment of inertia around the x-axis and y-axis, respectively; m_s represents the weight of the sprung mass; l is the half track-width; L_1 and

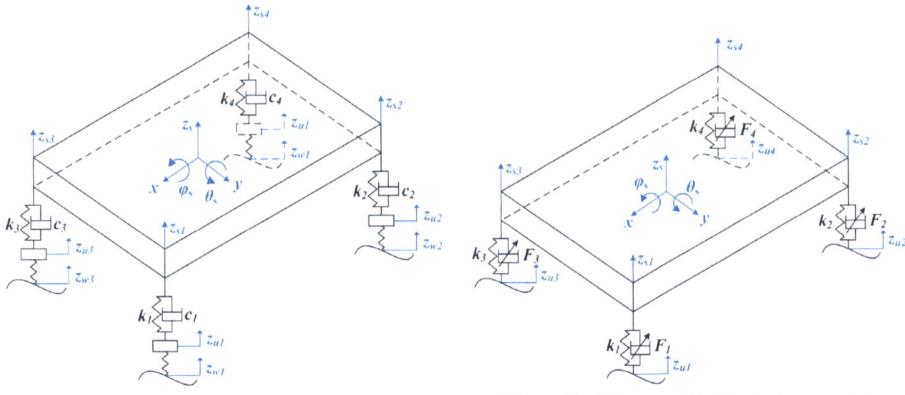

(a) Full car model with passive suspension　　(b) Simplified full car model with adaptive suspension

Fig. 2.4 Full-car models

L_2 are the distances from the vehicle CG location to the front and rear axles, respectively. Besides, F_{si} represents the spring forces, and F_i represents the damping forces as follows:

$$F_{si} = k_i(z_{ui} - z_{si})$$
$$F_i = c_i(\dot{z}_{ui} - \dot{z}_{si}) \tag{2.11}$$

The torques generated by the two anti-roll bars can be calculated by

$$\tau_{bar1} = k_{bar1}\left(\varphi_s - \frac{1}{2l}z_{u1} + \frac{1}{2l}z_{u3}\right)$$
$$\tau_{bar2} = k_{bar2}\left(\varphi_s - \frac{1}{2l}z_{u2} + \frac{1}{2l}z_{u4}\right) \tag{2.12}$$

Hence, the differential equations can be rewritten as:

$$m_s\ddot{z}_s = k_1(z_{u1} - z_{s1}) + k_2(z_{u2} - z_{s2}) + k_3(z_{u3} - z_{s3}) + k_4(z_{u4} - z_{s4}) + F_1 + F_2 + F_3 + F_4$$
$$I_x\ddot{\varphi}_s = k_1l(z_{u1} - z_{s1}) + k_2l(z_{u2} - z_{s2}) - k_3l(z_{u3} - z_{s3}) - k_4l(z_{u4} - z_{s4}) + F_1l + F_2l - F_3l - F_4l$$
$$- k_{bar1}\left(\varphi_s - \frac{1}{2l}z_{u1} + \frac{1}{2l}z_{u3}\right) - k_{bar2}\left(\varphi_s - \frac{1}{2l}z_{u2} + \frac{1}{2l}z_{u4}\right) + m_sa_yh$$
$$I_y\ddot{\theta}_s = -k_1L_1(z_{u1} - z_{s1}) + k_2L_2(z_{u2} - z_{s2}) - k_3L_1(z_{u3} - z_{s3}) + k_4L_2(z_{u4} - z_{s4})$$
$$- F_1L_1 + F_2L_2 - F_3L_1 + F_4L_2 - m_sa_xh \tag{2.13}$$

in which the suspension relative displacements can be simplified according to the small-angle simplifications:

$$z_{s1} = z_s + l\varphi_s - L_1\theta_s$$
$$z_{s2} = z_s + l\varphi_s + L_2\theta_s$$
$$z_{s3} = z_s - l\varphi_s - L_1\theta_s$$
$$z_{s4} = z_s - l\varphi_s + L_2\theta_s \tag{2.14}$$

The above equations can be written in the state-space form as follows:

$$\dot{X} = AX + B_uU + B_wW$$
$$Y = CX + D_uU + D_wW \tag{2.15}$$

in which the state vector X is

$$X = \begin{bmatrix} z_s & \varphi_s & \theta_s & \dot{z}_s & \dot{\varphi}_s & \dot{\theta}_s \end{bmatrix}^T \tag{2.16}$$

and the control input vector U is

$$U = \begin{bmatrix} F_1 & F_2 & F_3 & F_4 \end{bmatrix}^T \tag{2.17}$$

2.3 Full-Car Model

and the disturbance input vector W is

$$W = \begin{bmatrix} z_{u1} & z_{u2} & z_{u3} & z_{u4} & a_x & a_y \end{bmatrix}^T \qquad (2.18)$$

The full-car model is used to analyze various aspects of vehicle dynamics, including:

- **Ride Comfort**: Evaluating the vertical, pitch, and roll accelerations to ensure passenger comfort.
- **Handling and Stability**: Analyzing the vehicle's response to steering inputs and external disturbances to ensure optimal handling and stability.
- **Suspension Performance**: Assessing the ability of the suspension system to absorb road irregularities and maintain tire contact with the road.

Common Suspension Control Techniques   3

One of the key challenges in vehicle suspension control is achieving a balance between ride comfort and handling performance. While softer suspensions may provide a smoother ride, they can compromise vehicle stability and handling during aggressive driving maneuvers. Conversely, stiffer suspensions may offer improved handling at the expense of ride comfort. Effective suspension control strategies must strike a delicate balance between these competing objectives, optimizing suspension behavior based on driving conditions, vehicle dynamics, and driver preferences. Recent advancements in suspension control technology have led to the development of intelligent suspension systems capable of adapting to a wide range of driving scenarios. These systems employ predictive algorithms, adaptive damping mechanisms, and sensor feedback to continuously adjust suspension parameters based on real-time inputs. By dynamically optimizing suspension characteristics, intelligent suspension systems can deliver superior ride quality, enhanced vehicle stability, and improved traction across various driving conditions.

The Skyhook and Groundhook approaches were widely used in adaptive suspensions, and their hybrid algorithms were introduced in some studies [1]. For example, Liu et al. [2] introduced a generalized Skyhook-Groundhook (GenHook) hybrid strategy, in which the weighting coefficient of Skyhook and Groundhook in impedance-based control was determined via the 95th percentile aggregate fourth-power force. The simulation results demonstrated that the proposed controller could significantly enhance ride comfort and maintain stability under different working conditions. Past studies showed that Skyhook performed very well in minimizing the sprung mass' vertical acceleration around its resonance frequency but not in higher frequency ranges. However, the Acceleration Driven Damping (ADD) approach has complementary characteristics to Skyhook, which is good at isolating the vibrations beyond the sprung mass' natural frequency. Thus, the combination of these two comfort-oriented control algorithms was proposed in the literature, namely SH-ADD. It aimed to provide a quasi-optimal performance of ride quality. A crossover frequency α was selected near the first-order resonance frequency where SH and ADD were switched. In detail, the control algorithms of Skyhook, ADD, and SH-ADD are presented as follows [3]:

$$c_{SH} = \begin{cases} c_{min}, & \text{if } \dot{z}_s(\dot{z}_s - \dot{z}_u) < 0 \\ c_{max}, & \text{if } \dot{z}_s(\dot{z}_s - \dot{z}_u) \geq 0 \end{cases}$$

$$c_{ADD} = \begin{cases} c_{min}, & \text{if } \ddot{z}_s(\dot{z}_s - \dot{z}_u) < 0 \\ c_{max}, & \text{if } \ddot{z}_s(\dot{z}_s - \dot{z}_u) \geq 0 \end{cases}$$

$$c_{SH-ADD} = \begin{cases} c_{SH}, & \text{if } \ddot{z}_s^2 - \alpha^2 \dot{z}_s^2 < 0 \\ c_{ADD}, & \text{if } \ddot{z}_s^2 - \alpha^2 \dot{z}_s^2 \geq 0 \end{cases} \tag{3.19}$$

Moreover, Nie et al. [4] developed a modified SH-ADD with a frequency selector for a semi-active suspension to balance ride quality and road holding. The algorithm was validated through numerical analysis and experiment on a quarter-car test rig, which validated that the proposed controller could balance ride comfort and road holding around first- and second-order resonance frequencies. Besides, it reduced the chattering of the sprung mass acceleration, which was a common issue of the 'on–off' control algorithms.

In addition, H_2 and H_∞ control approaches were frequently mentioned in past studies, which included formal consideration of robustness with respect to model/parameter uncertainties [5]. Therefore, they were often referred to as robust controllers. Li et al. [6] developed an output feedback-based H_∞ controller for an active quarter-car suspension system. It was designed to ensure the asymptotic stability of the closed-loop system with H_∞ disturbance attenuation level and output constraints, and to improve ride comfort and road holding. Moreover, Du et al. [7] implemented a H_∞ optimal controller on a hydraulically interconnected suspension designed for three motion modes (bounce, pitch, and roll) using a common quadratic Lyapunov function. Li et al. [8] introduced a state-feedback H_2 controller for a quarter-car active suspension system with an adjustable inerter, a spring, and a damper in parallel. It aimed to improve ride quality, suspension deflection, and tire deflection performance. In the simulation, the inerter tracked the desired control force generated by the state feedback H_2 algorithm. The results demonstrated that the ride quality and tire deflection could be optimized with slight deterioration of the unsprung mass' natural frequency, and the suspension deflection was improved at high frequencies.

Some novel strategies were also introduced. For example, Zheng et al. [9] developed a novel framework of cloud-aided nonlinear active full car suspension, in which the backstepping control strategy was employed to handle the nonlinear characteristics updated in a remote cloud. The control reference was chosen based on the suspension aggregate information and road conditions, by which the control algorithm was processed on the central computational server. The uncertain parameters and unmodeled dynamics were dealt with by an adaptive algorithm. The simulation based on a 7 DOFs full car model demonstrated that the proposed strategy improved ride comfort by more than 80% compared to the passive suspension. Besides, Tang et al. [10] investigated a Takagi–Sugeno fuzzy controller for a semi-active quarter-car suspension equipped with an MR damper. The proposed approach was evaluated on a quarter-car test rig and compared to a Skyhook controller. The experimental results demonstrated that the TS fuzzy controller could effectively attenuate the undesired vibration by 10% to 30%. Moreover, Zhang et al. [11] introduced an adaptive neural network control scheme for an active suspension system by using bioinspired nonlinear dynamics. In detail, a novel constructive predictor was designed to address the input delay. Then, the neural networks were adopted to approximate the uncertain dynamics. A finite-time adaptive control was implemented to update the neural network weights and minimize vibrations. The theoretical analysis and experimental results indicated the efficiency of suppressing vibrations and low energy consumption by more than 44%. Additionally, Nguyen et al. [12] introduced a fractional-order derivative-based sliding mode control (FD-SMC) for a half-car semi-active suspension system with MR dampers, which aimed to minimize the sprung mass accelerations in vertical and pitch directions. The proposed controller consisted of I-SmDs, which relied on measured data processed by the new filter CoFilter [13] and the FD-based SMC technique. The FD-SMC determined the control force, and I-SmDs estimated the current intensity applied to MR dampers to generate the desired damping forces. The experimental test demonstrated that the proposed controller could deal with the dynamic response of the system and the noisy measurement very well. Specifically, the vertical and rotational accelerations of the sprung mass were reduced to a large extent and the road-holding ability was improved to provide safer driving conditions. An objective-oriented hierarchy control strategy that integrated with active braking and active front steering systems was developed by Lu et al. [14], in which a fuzzy control approach was used for the semi-active suspension system and a sliding mode control technique was utilized for braking and steering systems. The simulation results demonstrated that the proposed global integrated controller could effectively improve ride comfort, lateral stability, and braking safety. Coric et al. [15] introduced a collocation-type control variable optimization method to enhance vehicle ride comfort and maintain wheel-holding ability through a fully active suspension (FAS) system. A quarter-car and a full 10 DOFs vehicle models were used in this study. In addition, the cost function was extended to include the FAS energy consumption and wheel damage resilience costs. Na et al. [16] utilized a novel approximation-free control (AFC) technique to improve the efficiency and reliability of quarter-car active suspension systems. The major advantage of this approach was that it incorporated the unmolded dynamics of the hydraulic actuator and eliminated the need for laborious parameter tuning. Furthermore, the control strategy employed prescribed performance functions (PPFs) that ensured both transient and steady-state suspension performance for safe operations. The experiments based on a practical quarter-car active suspension test rig demonstrated that this control technique provided superior performance with improved computational efficiency.

3.1 Model-Based Control

Different model-based control techniques have been implemented in vehicle adaptive suspension systems. Among them, the MPC and LQR were the most popular algorithms in past studies. These two control approaches can solve multi-objective optimization problems, by which the vehicle dynamics and ride comfort in vertical, roll, and pitch directions can be analyzed and optimized. For example, Koch and Kloiber [17] introduced an adaptive controller which dynamically interpolated between differently tuned LQRs governed by the dynamics wheel load and the suspension deflection. Its performance was examined on a quarter-car test rig with a linear electrical motor. Compared to a fast adaptive Skyhook controller, the proposed approach could overcome conservatism and offer significant ride comfort while keeping the limits on the suspension deflection and the wheel load.

Compared to the LQR, the MPC algorithm can adequately handle the semi-active constraint, also known as the damping dissipativity constraint [18]. Nguyen et al. [19] designed an MPC controller for a semi-active suspension based on a 7 DOFs full car model. An observer that could estimate the road profile over the prediction horizon was introduced. This MPC controller with road estimation gave a trade-off between ride comfort and handling stability. Meanwhile, it guaranteed the physical constraints of semi-active dampers. Similar work has also been done by Gohrle [20]. However, the high computational cost of the MPC algorithm made it difficult and expensive to implement in practical applications. Therefore, various strategies have been proposed to maintain the accuracy of the prediction but reduce the computational cost at the same time, such as hybrid MPC [21], fast MPC [22], explicit MPC [23], nonlinear MPC [24], etc. Johan et al. [25] introduced an explicit MPC for an active suspension system. It was assessed through simulation and experiment on a sport utility vehicle (SUV). The simulation results showed that the vibrations in vertical, roll, and pitch directions could be reduced by 10% to 35% compared to the passive suspension. The experimental results validated that the proposed controller sufficiently minimized the vibration in the range of 0~15 Hz. However, this algorithm was processed on a dSPACE MicroAutoBox with 16MB flash memory which was more powerful than an automotive-grade microcontroller. Morato et al. [26] designed a fast real-time LPV MPC controller for a semi-active suspension with a sampling frequency of 200 Hz. The simulation results illustrated that the vertical vibrations were attenuated to a large extent, but the roll performance was at the same level as the passive suspension. The experimental test was done on a testbed called INOVE Soben-Car, but the algorithm was processed on a 2.4GHz 8GB RAM Macintosh computer instead of an automotive-grade microcontroller. Furthermore, a hybrid horizon-varying (HV) MPC controller with road preview was developed for autonomous vehicles on uneven roads [27]. This comfort optimization strategy combined vehicle speed planning and road-preview semi-active suspension control with a novel road data processing method, by which the control system could fully utilize the road information with a fixed preview length while adapting to speed variation. The simulation results verified the benefits of the hybrid HV-MPC method, especially in improving ride comfort and realizing multi-objective optimization (vertical vibration, driving time, and speed variation).

The damping forces were mostly considered as the control inputs in previous studies, which resulted in linear control plants with nonlinear state-dependent constraints. This issue increased the requirement of computational memory. Two different techniques were introduced by Kjellberg and Sundell [24] to solve this problem. Firstly, a full-complexity nonlinear MPC with 18 states was proposed. The full-car model was reformulated, which used damping coefficients as the control signal instead of damping forces. This method resulted in a nonlinear control plant with linear constraints. Secondly, the full car model was simplified by removing the wheel dynamics to reduce the execution time [28]. Hence, the number of states was reduced to 15. Meanwhile, the sampling time and prediction horizon were tuned to decrease the computational cost. The proposed real-time nonlinear MPC was finally implemented on a Volvo S90 equipped with semi-active dampers and a vision system, but the controller was run on a MicroAutoBox. Overall, it is noticeable that the real-time MPCs introduced in the past studies were not comprehensively examined on any automotive-grade microcontrollers. From the authors' point of view, the reasons can be concluded as follows:

- The QP solvers employed in most of the studies were ready-to-use packages/software, such as *quadprog* provided by MATLAB, *qpOASES* [29], *ODYS* [30], *NASOQ* from the University of Toronto [31], *OSQP* from the University of Oxford [32], etc. They usually require relatively high flash memories for processing, which causes a big problem for practical application.
- The prediction algorithm is an important feature of MPC, but it also needs a huge amount of flash memory. Meanwhile, the computational cost is highly dependent on the prediction horizon setting.
- Some studies focused on full car dynamics, resulting in a complicated mathematical model with more states that naturally increased the computational cost.

3.2 Gain-Adaptive Approach

Most existing works focused on non-adaptive control techniques, in which the controller gains were independent of the disturbance inputs. Since the driving conditions are randomly changed, using only one set of gains makes it impossible to achieve the optimum control performance all the time. Hence, some gain-adaptive algorithms are introduced in the literature, which can be classified into two main categories, i.e., parameter-adaptive suspension due to system nonlinear and uncertain properties [33] and road-adaptive suspension because of randomly changed road surfaces.

As one of the most commonly used suspension control techniques, Skyhook control was widely utilized for developing adaptive control systems in the literature. Nguyen et al. [34] designed a road frequency adaptive control system for a quarter-car semi-active suspension to improve both ride comfort and handling performances in all frequency ranges of road disturbances. Their control law was extended from the conventional Skyhook approach, and its gains were scheduled based on the estimated disturbance frequencies. To determine the gains of the modified Skyhook algorithm, a fully active LQR controller was utilized to generate the optimum damping force (benchmark). The scheduled gains aimed to minimize the force differences between the modified Skyhook and the active LQR using the Minimum Norm Criterion [35]. The performance of the proposed controller was evaluated through numerical simulations and only compared to the passive suspension system, which showed that the adaptive Skyhook could provide better ride comfort and handling ability. Similarly, Hong et al. [36] developed a road-adaptive Skyhook controller for a semi-active Macpherson suspension. The proposed approach extended the conventional Skyhook-Groundhook control scheme and scheduled its gains according to road conditions based on the measured sprung mass' vertical acceleration. Five sets of optimal gains were presented for different ISO roads from class A (very good) to class E (very poor) to describe the road roughness. The hardware-in-the-loop simulations examined the control algorithm on a quarter-car test rig, demonstrating that the proposed semi-active suspension achieved a competitive control performance by adopting the road-adaptive law. Moreover, Yi and Song [37] developed a road-adaptive Skyhook controller using a Neural Network based on a road detection algorithm. The modified Skyhook approach considered both ride quality and dynamic tire force. The proposed road detection algorithm determined and updated the Skyhook gains according to the estimated frequency contents of road disturbances based on the measured sprung mass acceleration. Its performance was validated through a quarter-car test rig. The experimental results showed that the proposed road-adaptive Skyhook controller provided adequate damping for the wheel hop frequency and enhanced the ride comfort and road holding compared to the classic Skyhook approach.

Besides the Skyhook approach, other control techniques were utilized to achieve road-adaptive functionality. Koch et al. [38] designed a road-adaptive LQR controller for a quarter-car active suspension system. With a switching control structure, the adaptive algorithm scheduled the LQR gains according to the dynamic wheel load and the suspension deflection. Specifically, there were six pre-defined sets of LQR weighting matrices for different primary control objectives, such as comfort-oriented, intermediate, safety-oriented, etc. The switching logic switched between six sets of weighting matrices to maximize the overall suspension performance. The simulation results showed that the proposed road-adaptive LQR controller improved the ride comfort by 20% compared to the passive suspension and 11% compared to the non-adaptive LQR approach.

3.2 Gain-Adaptive Approach

Similarly, Fialho and Balas [39] presented a framework for developing a road-adaptive suspension controller via LPV control and nonlinear back-stepping techniques. The gains of the LPV algorithm were adaptively switched between two pre-tuned values based on the measured suspension deflections, and a first-order filter was implemented to avoid discontinuous gain transition.

Based on the review of past studies, some research topics on vehicle suspension systems can be investigated:

- **Coordinated vibration attenuation**: The majority of past studies aimed to reduce the vehicle's vertical acceleration and, to a lesser extent, pitch vibrations. However, the tilt motions in roll and pitch directions are also the primary sources of discomfort. As such, damping control systems that can improve vehicle dynamics in all three directions should be developed.
- **Advanced semi-active suspension systems**: A semi-active suspension system requires less power than an active one, but the research on addressing the damper dissipative constraint and the mechanical limitation should be thoroughly considered. Meanwhile, the control algorithm needs to be concise enough to be processed on an automotive-grade microcontroller.
- **Gain-adaptive control approach**: Vehicles may drive on various kinds of roads, such as highways, rough paved roads, off-roads, etc. It is not possible to find one set of gains that can handle a variety of driving conditions. One potential solution is to intelligently adjust the control parameters (such as objective functions, penalty matrices, gains, etc.) according to the disturbances. This approach requires road classifiers to provide accurate disturbance information for the control system. In addition, proper gain-tuning rules need to be developed to guarantee robust performances on any road.

4 Suspension Control Applications

In this chapter, the Skyhook, LQR, MPC, and an integrated Skyhook-LQR algorithms are thoroughly introduced, with detailed implementations for sedan, SUV, and truck systems. Multiple examples simulated in MATLAB/Simulink and CarSim environments are provided to help the readers understand and replicate these algorithms.

4.1 Skyhook Control

In the early 1970s, Karnopp and Crosby first introduced the concept of the semi-active suspension system, based on the well-known Skyhook control [40]. Because of the intuitive approach and good performance, it is widely used in (semi-)active suspension systems. This comfort-oriented control law focuses on reducing the absolute sprung mass acceleration, by generating a damping force proportional to the absolute sprung mass velocity. The Skyhook algorithm is a classical control strategy for vehicle suspension systems, which mainly focuses on vertical vibration attenuation. The concept of the Skyhook control is that the sprung mass connects to an imaginary sky through a damper, and a spring installs between the sprung mass and unsprung mass, as shown in Fig. 4.1.

This approach in the literature consists of two control states, i.e., low-damped and high-damped. The damping coefficient c_{sky} changes according to the sign of the product of the sprung mass and unsprung mass velocities as follows:

$$\begin{cases} c_{sky} = c_{max}, & if \, \dot{z}_s(\dot{z}_s - \dot{z}_f) \geq 0 \\ c_{sky} = c_{min}, & if \, \dot{z}_s(\dot{z}_s - \dot{z}_f) < 0 \end{cases} \quad (4.1)$$

Fig. 4.1 Skyhook control approach

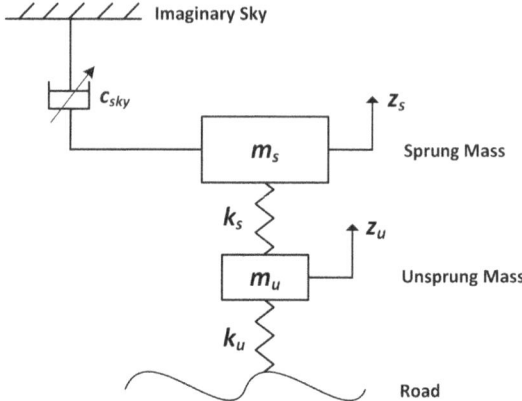

It can be seen that the damping coefficient c_{sky} switches between the maximum valve c_{max} and the minimum value c_{min}. In the active continuous Skyhook control strategy, if the product $\dot{z}_s(\dot{z}_s - \dot{z}_f)$ is greater than or equal to zero, the high state is applied. If the product is negative, the damper is adjusted to the low state. After determining the damping coefficient, the corresponding control signal F_{sky} is calculated by

$$F_{sky} = -c_{sky}\dot{z}_s \tag{4.2}$$

Skyhook control is a powerful strategy for vehicle suspension systems, offering significant improvements in ride comfort. By intelligently adjusting the damping force or applying active control forces, Skyhook control effectively counteracts body movements, resulting in a smoother and more controlled ride. It has been widely used in modern vehicles, particularly in high-end cars.

Example 4.1 Develop a conventional Skyhook controller for a semi-active suspension system in MATLAB, using the following parameters of a mid-sized SUV:

```
Ms = 2257;      % vehicle sprung mass
Ix = 846.6;     % roll moment of inertia
Iy = 3524.9;    % pitch moment of inertia
L1 = 1.33;      % CG to front axle
L2 = 1.725;     % CG to rear axle
l = 0.8625;     % half of track width
h = 0.781;      % CG height
kbar1 = 32601;  % roll stiffness of front ARB
kbar2 = 29221;  % roll stiffness of rear ARB
k1 = 1.89e5;    % front left spring stiffness
k2 = 62.7e3;    % rear left spring stiffness
k3 = 1.89e5;    % front right spring stiffness
k4 = 62.7e3;    % rear right spring stiffness
```

4.1 Skyhook Control

The two rear spring stiffness can be modelled using a 1-D Lookup Table in Simulink, as shown in Fig. 4.2.

The Skyhook controller reads the vehicle CG motions in vertical, roll, and pitch directions from a 6-axis IMU. The movements at the four corners of the vehicle can be calculated as follows:

```
zs1_dot = zs_velocity + l*roll_rate - L1*pitch_rate;
zs2_dot = zs_velocity + l*roll_rate + L2*pitch_rate;
zs3_dot = zs_velocity - l*roll_rate - L1*pitch_rate;
zs4_dot = zs_velocity - l*roll_rate + L2*pitch_rate;
zs_dot = [zs1_dot; zs2_dot; zs3_dot; zs4_dot];
```

Then, using the Skyhook algorithm to calculate the adaptive damping forces:

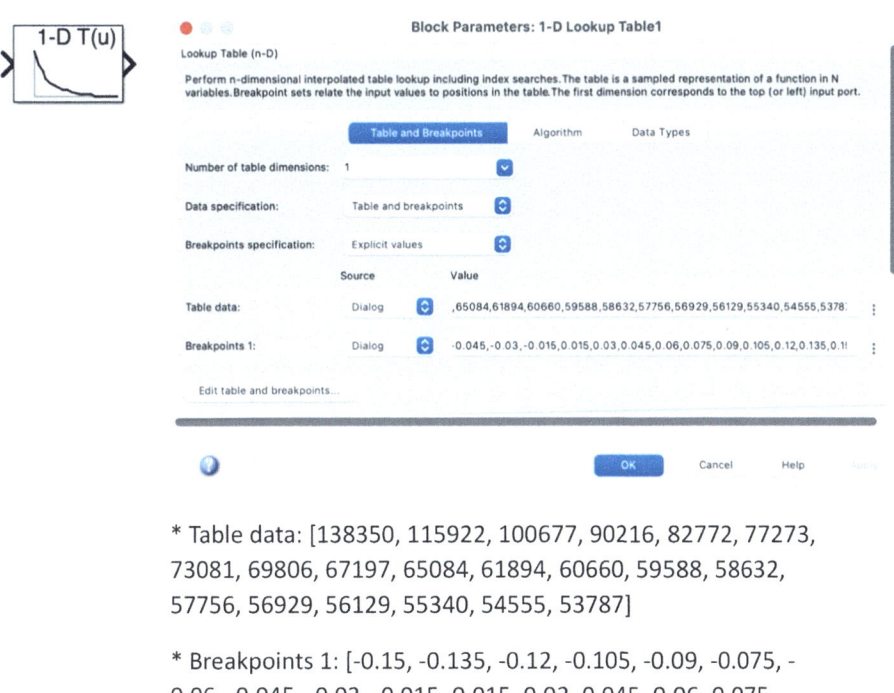

* Table data: [138350, 115922, 100677, 90216, 82772, 77273, 73081, 69806, 67197, 65084, 61894, 60660, 59588, 58632, 57756, 56929, 56129, 55340, 54555, 53787]

* Breakpoints 1: [-0.15, -0.135, -0.12, -0.105, -0.09, -0.075, -0.06, -0.045, -0.03, -0.015, 0.015, 0.03, 0.045, 0.06, 0.075, 0.09, 0.105, 0.12, 0.135, 0.15]

Fig. 4.2 2D Lookup Table used for modelling air spring stiffness in MATLAB/Simulink

```
F_min = zeros(4,1);
F_max = zeros(4,1);
c_nom = 5e3;     % needs to be fine-tuned
F_sky = zeros(4,1);    % imaginary damping force
c_sky = zeros(4,1);    % imaginary damper coefficient
c_damper = zeros(4,1); % real damper coefficient
F_damper = zeros(4,1); % real damping force
for i = 1:4     % CDC damper
    delta_v = -rel_susp_vel(i);   % lower - upper
    if delta_v >= 0 && delta_v<= 0.262 % Compression
        F_min(i) = 839 * delta_v;
        F_max(i) = 5466 * delta_v;
    elseif delta_v > 0.262
        F_min(i) = 839 * delta_v;
        F_max(i) = 1226 * delta_v + 1111;
    elseif delta_v < 0 && delta_v >= -0.262 % Rebound
        F_min(i) = 13948 * delta_v;
        F_max(i) = 1381 * delta_v;
    elseif delta_v < -0.262
        F_min(i) = 1438 * delta_v - 3278;
        F_max(i) = 1381 * delta_v;
    end


    if -delta_v * zs_dot(i) >=0     % high state
        c_sky(i) = c_nom;
    else   % low state
        c_sky(i) = 0;
    end
    F_sky(i) = -c_sky(i) * zs_dot(i);
    if F_sky(i) > F_max(i)
        F_damper(i) = F_max(i);
    elseif F_sky(i) < F_min(i)
        F_damper(i) = F_min(i);
    else
        F_damper(i) = F_sky(i);
    end
end
```

in which the F_damper is the output damping forces given by the Skyhook algorithm.

The offset-bump road profile used in the simulation is shown in Fig. 4.3 which is also known as the Twist Ditch. As can be seen, it consists of two offset rounded bumps with 3 inches in height. Each bump is offset 10 feet peak to peak from left to right. The vehicle speed is 18 km/h.

The acceleration responses at the vehicle CG location in vertical, roll, and pitch directions are presented in Figs. 4.4, 4.5, and 4.6, respectively. The vertical vibration is significantly reduced by Skyhook with a 26% improvement. Similarly, the reduction of pitch acceleration is satisfactory, with 16% by Skyhook. Besides, the roll vibration was slightly attenuated by 8%. The damping forces requested by the Skyhook controller are presented in Fig. 4.7, which are within the practical boundary.

As a comfort-oriented control approach, Skyhook focuses on reducing the absolute vertical acceleration of the sprung mass. Hence, the Skyhook approach has been widely used

Fig. 4.3 Offset-bump road profile

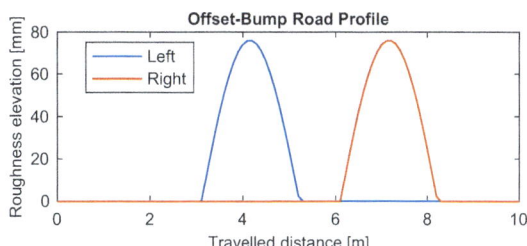

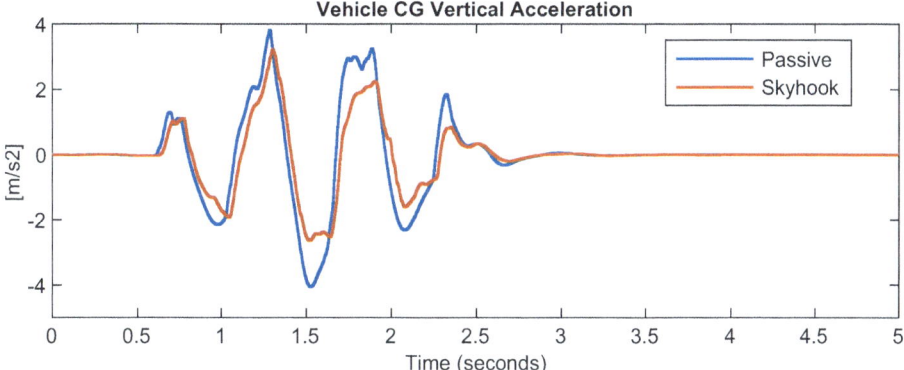

Fig. 4.4 Vehicle vertical acceleration at CG location

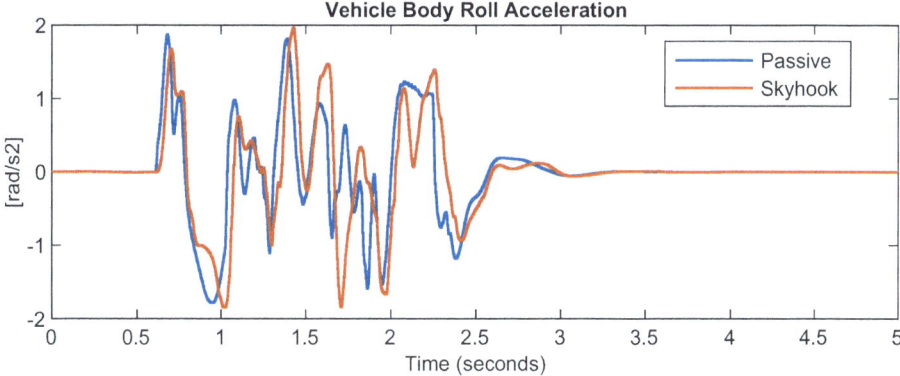

Fig. 4.5 Vehicle roll acceleration

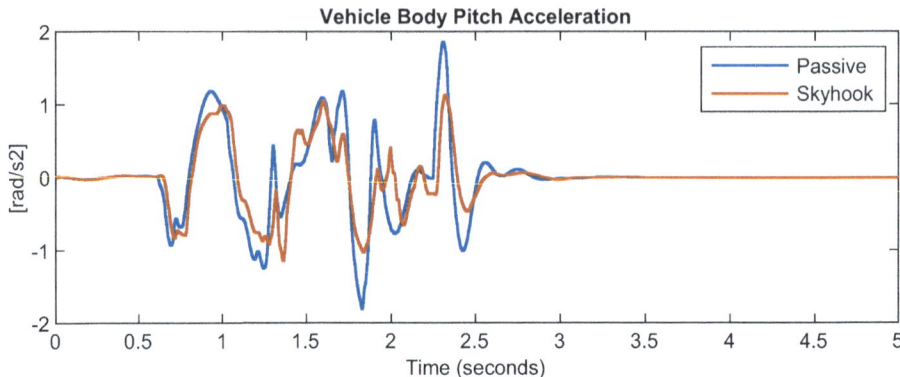

Fig. 4.6 Vehicle pitch acceleration

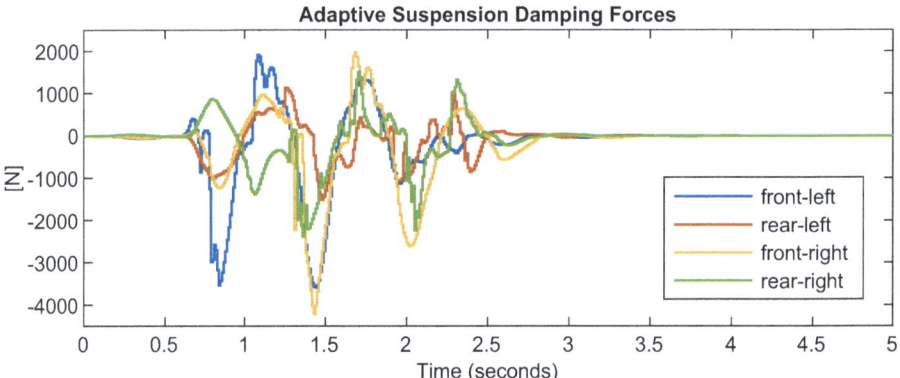

Fig. 4.7 Adaptive damping forces (operated at 50Hz)

in minimizing vertical vibrations. However, the conventional Skyhook control is not always adequate for optimizing the vehicle's rotational motions. To fill this gap, the LQR control approach can be a solution.

4.2 Linear Quadratic Regulator (LQR)

Linear Quadratic Regulator (LQR) is a method in control theory to design a controller that regulates the behavior of a dynamic system to achieve optimal performance. It involves minimizing a cost function, which is typically quadratic in terms of state variables and

4.2 Linear Quadratic Regulator (LQR)

control inputs. The goal is to determine a state feedback law that minimizes the cumulative cost, balancing performance and control effort. LQR is widely used in vehicle control systems due to its optimal performance and ability to handle multi-variable control problems, such as adaptive suspension control, vehicle stability control, path following and lane keeping, adaptive cruise control, etc.

For the proposed linear time-invariant vehicle suspension system introduced in Eq. 2.15 in Chap. 2, the cost function J is firstly defined in a quadratic form shown in Eq. 4.3. It aims to optimize vehicle dynamics in vertical, pitch, and roll directions.

$$J = \int_0^\infty Y^T Q_y Y + U^T R_y U dt \qquad (4.3)$$

in which the output penalty matrix $Q_y \geq 0$, the control signal weighting matrix $R_y > 0$, and the output vector Y is defined as

$$Y = \dot{X} = \begin{bmatrix} \dot{z}_s & \dot{\theta}_s & \dot{\varphi}_s & \ddot{z}_s & \ddot{\theta}_s & \ddot{\varphi}_s \end{bmatrix}^T \qquad (4.4)$$

The above cost function needs to be transformed into the standard form as follows:

$$J = \int_0^\infty X^T \underbrace{C^T Q_y C}_{Q} X + U^T \underbrace{\left(R_y + D^T Q_y D \right)}_{R} U \, dt \qquad (4.5)$$

The feedback control law that minimizes the above cost function is

$$U = -KX \qquad (4.6)$$

in which the feedback gain K can be calculated by

$$K = R^{-1} B^T P \qquad (4.7)$$

where P can be solved from the continuous-time algebraic Riccati equation (ARE) as follows:

$$A^T P + PA - PBR^{-1} B^T P + Q = 0 \qquad (4.8)$$

The ARE is the key to solving optimal control problems, especially in the design of the LQR controller. The equation arises in the context of determining the optimal state feedback control law for a linear system to minimize a quadratic cost function. In MATLAB, the continuous-time ARE can be solved using the `care` function. The solution to the ARE ensures that the control system achieves the desired performance while maintaining stability and efficiency.

Example 4.2 Develop an LQR controller for an active suspension system in MATLAB, using the following parameters of a high-end sedan:

```
Ms = 2164;      % vehicle sprung mass
Ix = 928.1;     % roll moment of inertia
Iy = 2788.5;    % pitch moment of inertia
L1 = 1.4;       % CG to front axle
L2 = 1.66;      % CG to rear axle
l = 0.8;        % half of track width
h = 0.53;       % CG height
kbar1 = 22002;  % roll stiffness of front ARB
kbar2 = 19710;  % roll stiffness of rear ARB
k1 = 46e3;      % front-left spring stiffness
k2 = 62.7e3;    % rear-left spring stiffness
k3 = 46e3;      % front-right spring stiffness
k4 = 62.7e3;    % rear-right spring stiffness
```

Step 1: System modelling using the simplified full-car state-space formulation in Eq. 2.15.

Step 2: Determine LQR weighting matrices. The Q and R matrices are chosen based on design specifications, which usually need to be fine-tuned. In this example, the following gains are chosen:

```
Qy = diag([0 0 0 7e3 7e3 7e3]);
Ry = 1e-3 * eye(4);
```

Step 3: Use the *lqr* function in MATLAB to calculate the feedback gain K.

```
Q = C' * Qy * C;
R = Du' * Qy * Du + Ry;
K_lqr = lqr(A, Bu, Q, R);
F = -K_lqr*X; % Output damping forces
```

Step 4: Set up the co-simulation environment of MATLAB/Simulink and CarSim for high-fidelity simulation studies, as shown in Fig. 4.8. CarSim is a sophisticated simulation software widely used in the automotive industry and academia for vehicle dynamics analysis and development. It provides detailed, physics-based models for simulating the behavior of vehicles, allowing engineers and researchers to test and evaluate vehicle performance, handling, and control systems under various driving conditions. CarSim includes comprehensive tools for modeling and simulating everything from simple passenger cars to complex multi-vehicle scenarios. It integrates seamlessly with other engineering software and hardware, facilitating the design and optimization of advanced vehicle systems.

The offset-bump road profile used in this scenario is shown in Fig. 4.3, and the vehicle speed is 25 km/h. The acceleration responses at the vehicle CG location in vertical, roll,

4.2 Linear Quadratic Regulator (LQR)

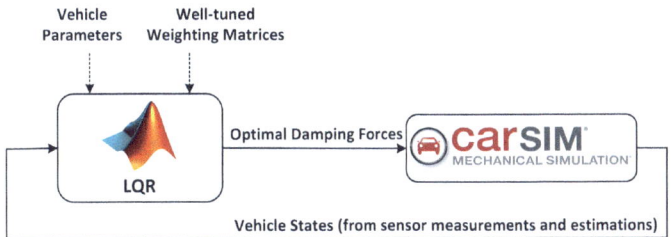

Fig. 4.8 A general development structure of a LQR suspension controller in MATLAB/Simulink and CarSim co-simulation environment

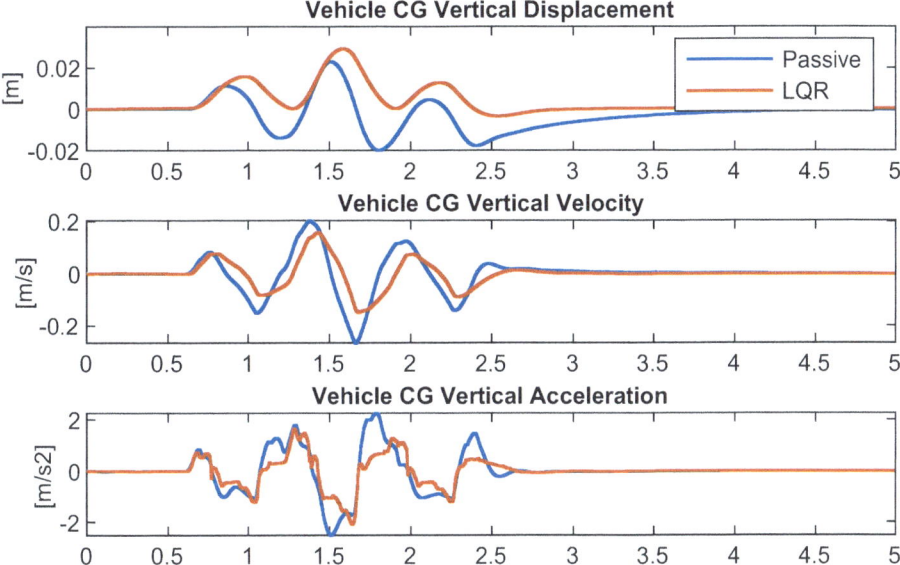

Fig. 4.9 Vehicle vertical responses at CG location

and pitch directions are presented in Figs. 4.9, 4.10, and 4.11, respectively. In the vertical direction, the LQR controller reduces the undesired vibrations by 31% in terms of the acceleration's RMS value. Its performance in roll direction is also significant, providing a 35% improvement. Similarly, the proposed LQR controller reduces the pitch acceleration by 34%. The requested damping forces are presented in Fig. 4.12, which are within the practical boundary.

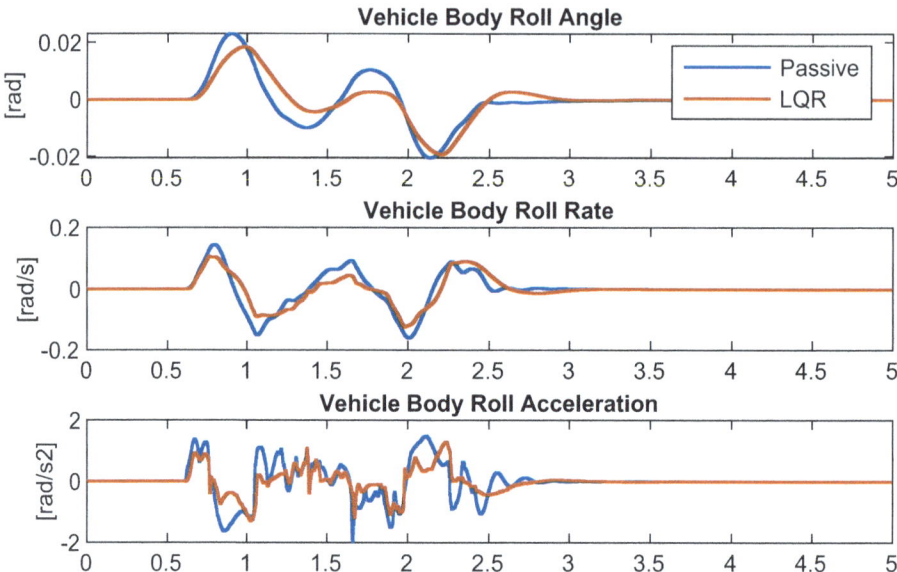

Fig. 4.10 Vehicle body motions in roll direction

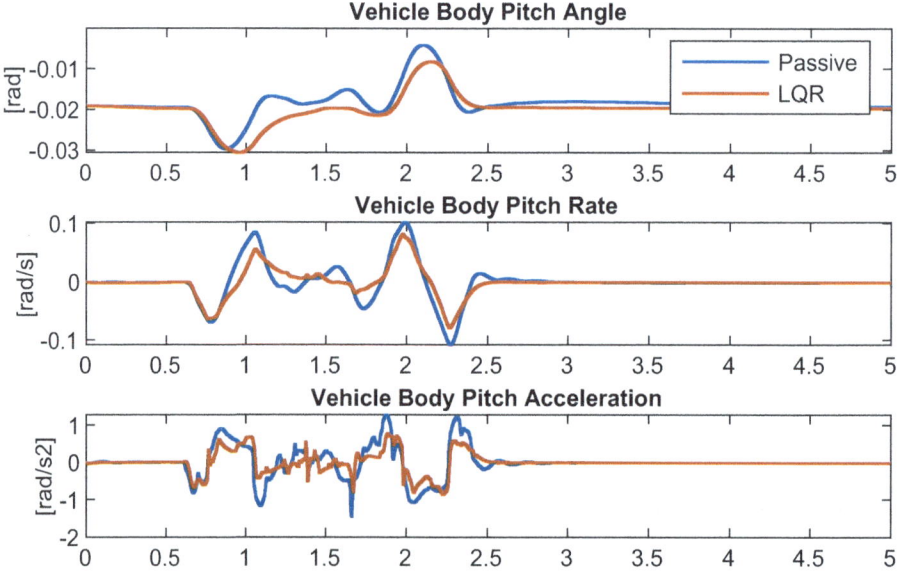

Fig. 4.11 Vehicle body motions in pitch direction

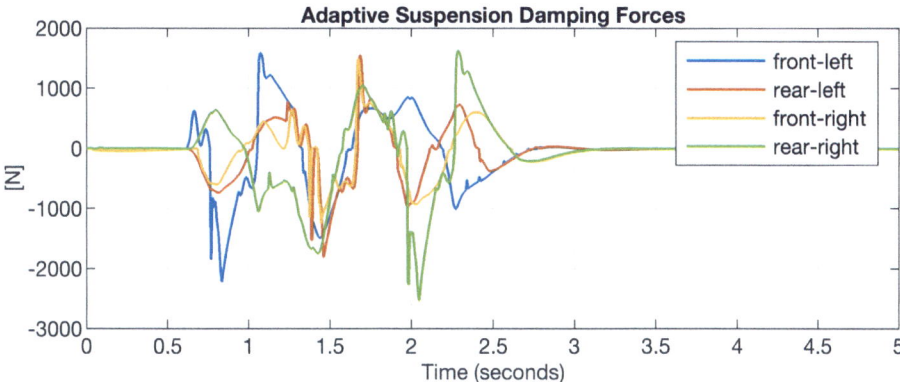

Fig. 4.12 Adaptive damping forces

4.3 Model Predictive Control (MPC)

Considering the multi-objective semi-active control problem with constraints, the Model Predictive Control has gained more attention in recent decades. MPC is a model-based dynamic optimizer, which is naturally capable of handling multi-variable systems. Furthermore, the state and input constraints can be easily formulated into the optimization problem. Besides, MPC has inherent local robustness to disturbances and uncertainties, as it can be designed to locally recover the LQR behavior [41]. The system mathematical model is used to predict the future states over a certain horizon, and the control action sequence is achieved by repeatedly solving finite-time optimization problems in a receding horizon fashion [42]. Specifically, the optimization is formulated as a constrained quadratic programming problem. The core of MPC is to utilize the receding horizon idea, its principle is presented as follows [24]:

(1) At time instant k, based on the current states of the control plant, the controller uses a mathematical model to predict the model output over a finite prediction horizon N_p.
(2) The optimization problem is solved over the control horizon N_c based on a predefined cost function subject to the system constraints. An optimal control sequence is generated in this step.
(3) The first element in the control sequence is applied to the system at the current time instant, and the rest elements are discarded.
(4) Return to step 1 for the next time instant $k + 1$.

The MPC controller is designed to predict the vehicle's vertical, roll, and pitch motions with the purpose of improving its ride comfort and stability. The prediction model's accuracy and complexity have a critical impact on the closed-loop optimization, i.e., computational cost and control performance. The prediction model of the vehicle system is explicitly given in a state-space formulation as introduced in Eq. 2.15. The proposed linear-time-invariant (LTI) mathematical model is discretized by using the Euler method with a sampling period T_s. Then, the discretized matrices can be written as follows:

$$A_d = I + T_s A$$
$$B_{ud} = T_s B_u$$
$$B_{wd} = T_s B_w$$
$$C_d = C$$
$$D_{ud} = D_u$$
$$D_{wd} = D_w \quad (4.9)$$

and the discretized plant model can be formed as:

$$X(k+1) = A_d X(k) + B_{ud} U(k) + B_{wd} W(k)$$
$$Y(k) = C_d X(k) + D_{ud} U(k) + D_{wd} W(k) \quad (4.10)$$

where A_d is the state transition matrix, B_{ud} is the control input matrix, B_{wd} is the disturbance input matrix, C_d is the output matrix, and D_{ud} and D_{wd} are feedforward matrices. The subscript d is used to clarify that the matrices are discrete, which are from now on neglected and the matrices are assumed to be discrete.

The state predictions over the prediction horizon are:

$$\hat{X}(k+1|k) = A\hat{X}(k|k) + B_u U(k|k) + B_w W(k|k)$$

$$\begin{aligned}\hat{X}(k+2|k) &= A\hat{X}(k+1|k) + B_u U(k+1|k) + B_w W(k+1|k) \\ &= A\left[A\hat{X}(k|k) + B_u U(k|k) + B_w W(k|k)\right] + B_u U(k+1|k) + B_w W(k+1|k) \\ &= A^2 \hat{X}(k|k) + AB_u U(k|k) + AB_w W(k|k) + B_u U(k+1|k) + B_w W(k+1|k)\end{aligned} \quad (4.11)$$

$$\vdots$$

$$\begin{aligned}\hat{X}(k+N_p|k) &= A\hat{X}(k+N_p-1|k) + B_u U(k+N_c-1|k) + B_w W(k+N_c-1|k) \\ &= A^{N_p} \hat{X}(k|k) + A^{N_c-1} B_u U(k|k) + A^{N_c-1} B_w W(k|k) \\ &\quad + A^{N_c-2} B_u U(k+1|k) + A^{N_c-2} B_w W(k+1|k) + \cdots \\ &\quad + B_u U(k+N_c-1|k) + B_w W(k+N_c-1|k)\end{aligned}$$

in which N_p is the prediction horizon, and N_c is the control horizon.

The output predictions over the prediction horizon are:

$$\hat{Y}(k|k) = C\hat{X}(k|k) + D_u U(k|k) + D_w W(k|k)$$

4.3 Model Predictive Control (MPC)

$$\begin{aligned}\hat{Y}(k+1|k) &= C\big[A\hat{X}(k|k) + B_u U(k|k) + B_w W(k|k)\big] + D_u U(k+1|k) + D_w W(k+1|k)\\&= CA\hat{X}(k|k) + CB_u U(k|k) + CB_w W(k|k) + D_u U(k+1|k) + D_w W(k+1|k)\end{aligned} \quad (4.12)$$

$$\vdots$$

$$\begin{aligned}\hat{Y}(k+N_p|k) &= C\hat{X}(k+N_p|k) + D_u U(k+N_p|k) + D_w W(k+N_p|k)\\&= CA^{N_p}\hat{X}(k|k) + CA^{N_p-1}B_u U(k|k) + CA^{N_p-1}B_w W(k|k)\\&\quad + CA^{N_p-2}B_u U(k+1|k) + CA^{N_p-2}B_w W(k+1|k) + \cdots\\&\quad + CA^{N_p-N_c}B_u U(k+N_c-1|k) + CA^{N_p-N_c}B_w W(k+N_c-1|k)\\&\quad + CB_u U(k+N_c|k) + CB_w W(k+N_c|k)\\&\quad + D_u U(k+N_c|k) + D_w W(k+N_c|k)\end{aligned}$$

Hence, the state predictions can be rewritten into the matrix form:

$$\begin{aligned}\begin{bmatrix}\hat{X}(k+1|k)\\\hat{X}(k+2|k)\\\vdots\\\hat{X}(k+N_p-1|k)\\\hat{X}(k+N_p|k)\end{bmatrix} &= \begin{bmatrix}A\\A^2\\\vdots\\A^{N_p-1}\\A^{N_p}\end{bmatrix}\hat{X}(k|k)\\&\quad + \begin{bmatrix}B_u & 0 & 0 & \cdots & 0\\AB_u & B_u & 0 & \ddots & 0\\\vdots & \ddots & \ddots & \ddots & 0\\A^{N_p-2}B_u & A^{N_p-3}B_u & \cdots & A^{N_p-N_c-2}B_u & 0\\A^{N_p-1}B_u & A^{N_p-2}B_u & \cdots & A^{N_p-N_c-1}B_u & A^{N_p-N_c}B_u\end{bmatrix}\begin{bmatrix}U(k|k)\\U(k+1|k)\\\vdots\\U(k+N_c-2|k)\\U(k+N_c-1|k)\end{bmatrix}\\&\quad + \begin{bmatrix}B_w & 0 & 0 & \cdots & 0\\B_w & B_w & 0 & \ddots & 0\\\vdots & \ddots & \ddots & \ddots & 0\\B_w & B_w & \ddots & B_w & 0\\B_w & B_w & \cdots & B_w & B_w\end{bmatrix}\begin{bmatrix}W(k|k)\\W(k+1|k)\\\vdots\\W(k+N_c-2|k)\\W(k+N_c-1|k)\end{bmatrix}\end{aligned} \quad (4.13)$$

Example 4.3 Implement the above matrices in MATLAB.

```
Np = 5;   % prediction horizon
Nc = 5;   % control horizon
Q_MPC = zeros(6*Np,6*Np);
q = diag([0 0 0 1e30 0]);
for i=1:Np
    Q_MPC(6*(i-1)+1:6*i,6*(i-1)+1:6*i) = q;
end

R_MPC = 1e-3 * eye(4*(Nc+1),4*(Nc+1));

%% Input Predictions: X_prediction = SA*X + Su*Fd + Sw*W
% SA
SA = zeros(6*Np,6);
for I = 1:Np
    SA(6*(i-1)+1:6*i,1:6) = Ad^i;
end
% Su
Su = zeros(6*Np,4*Nc);
for row = 1:Np
    for column = 1:Nc
        if column <= row
            Su(6*(row-1)+1:6*row, 4*(column-1)+1:4*column) = Ad^(row-column) * Bud;
        else
            Su(6*(row-1)+1:6*row, 4*(column-1)+1:4*column) = zeros(6,4);
        end
    end
end
% Sw
Sw = zeros(6*Np,6*Nc);
for row = 1:Np
    for column = 1:Nc
        if column <= row
            Sw(6*(row-1)+1:6*row, 6*(column-1)+1:6*column) = Bwd;
        else
            Sw(6*(row-1)+1:6*row, 6*(column-1)+1:6*column) = zeros(6,6);
        end
    end
end
```

The output predictions can be rewritten into matrix form as:

$$\underbrace{\begin{bmatrix} \hat{Y}(k|k) \\ \hat{Y}(k+1|k) \\ \vdots \\ \hat{Y}(k+N_p-2|k) \\ \hat{Y}(k+N_p-1|k) \end{bmatrix}}_{\hat{Y}} = \underbrace{\begin{bmatrix} C \\ CA \\ \vdots \\ CA^{N_p} \\ CA^{N_p-1} \end{bmatrix}}_{\Phi} \hat{X}(k|k)$$

$$+ \underbrace{\begin{bmatrix} D_u & 0 & 0 & \cdots & 0 \\ CB_u & D_u & 0 & \ddots & 0 \\ \vdots & \ddots & \ddots & \ddots & 0 \\ CA^{N_p-3}B_u & CA^{N_p-4}B_u & \ddots & D_u & 0 \\ CA^{N_p-2}B_u & CA^{N_p-3}B_u & \cdots & CA^{N_p-N_c}B_u & D_u \end{bmatrix}}_{\Gamma_u} \underbrace{\begin{bmatrix} U(k|k) \\ U(k+1|k) \\ \vdots \\ U(k+N_c-2|k) \\ U(k+N_c-1|k) \end{bmatrix}}_{\hat{U}}$$

4.3 Model Predictive Control (MPC)

$$+ \underbrace{\begin{bmatrix} D_w & 0 & 0 & \cdots & 0 \\ CB_w & D_w & 0 & \ddots & 0 \\ \vdots & \ddots & \ddots & \ddots & 0 \\ CA^{N_p-3}B_w & CA^{N_p-4}B_w & \ddots & D_w & 0 \\ CA^{N_p-2}B_w & CA^{N_p-3}B_w & \cdots & CB_w & D_w \end{bmatrix}}_{\Gamma_w} \underbrace{\begin{bmatrix} W(k|k) \\ W(k+1|k) \\ \vdots \\ W(k+N_c-2|k) \\ W(k+N_c-1|k) \end{bmatrix}}_{\hat{W}} \qquad (4.14)$$

Example 4.4 Implement the above matrices in MATLAB.

```
% PHI
PHI = zeros(6*Np,6);
for i=1:Np
    PHI(6*(i-1)+1:6*i,1:6) = Cd * Ad^i;
end

% Tu
Tu = zeros(6*Np,4*(Nc+1));
for row = 1:Np
    for column = 1:Nc+1
        if column <= row
            Tu(6*(row-1)+1:6*row, 4*(column-1)+1:4*column) = Cd * Ad^(row-column) * Bud;
        elseif column == row+1
            Tu(6*(row-1)+1:6*row, 4*(column-1)+1:4*column) = Dud;
        else
            Tu(6*(row-1)+1:6*row, 4*(column-1)+1:4*column) = zeros(6,4);
        end
    end
end

% Tw
Tw = zeros(6*Np,6*(Nc+1));


for row = 1:Np
    for column = 1:Nc+1
        if column <= row
            Tw(6*(row-1)+1:6*row, 6*(column-1)+1:6*column) = Cd * Ad^(row-column) * Bwd;
        elseif column == row+1
            Tw(6*(row-1)+1:6*row, 6*(column-1)+1:6*column) = Dwd;
        else
            Tw(6*(row-1)+1:6*row, 6*(column-1)+1:6*column) = zeros(6,6);
        end
    end
end
```

The above equation can be written short as:

$$\hat{Y} = \Phi \hat{X}(k) + \Gamma_u \hat{U} + \Gamma_w \hat{W} \quad (4.15)$$

In order to minimize the output of the system, a quadratic optimization problem is formulated. The objectives are minimizing the vehicle's vertical, roll, and pitch accelerations to improve ride quality. The objective function J is written in the Quadratic Programming (QP) problem as follows:

$$J = \min_{\hat{U}} \hat{Y}^T Q \hat{Y} + \hat{U}^T R \hat{U} \quad (4.16)$$

where Q is the output penalty matrix and R is the control signal penalty matrix.

By substituting output prediction $\hat{Y}$ into J, the following objective function is derived:

$$\begin{aligned}
\hat{Y}^T Q \hat{Y} + \hat{U}^T R \hat{U} &= \left[\Phi \hat{X}(k) + \Gamma_u \hat{U} + \Gamma_w \hat{W}\right]^T Q \left[\Phi \hat{X}(k) + \Gamma_u \hat{U} + \Gamma_w \hat{W}\right] + \hat{U}^T R \hat{U} \\
&= \hat{X}(k)^T \Phi^T Q \Phi \hat{X}(k) + \hat{X}(k)^T \Phi^T Q \Gamma_u \hat{U} + \hat{X}(k)^T \Phi^T Q \Gamma_w \hat{W} \\
&\quad + \hat{U}^T \Gamma_u^T Q \Phi \hat{X}(k) + \hat{U}^T \Gamma_u^T Q \Gamma_u \hat{U} + \hat{U}^T \Gamma_u^T Q \Gamma_w \hat{W} \\
&\quad + \hat{W}^T \Gamma_w^T Q \Phi \hat{X}(k) + \hat{W}^T \Gamma_w^T Q \Gamma_u \hat{U} + \hat{W}^T \Gamma_w^T Q \Gamma_w \hat{W} + \hat{U}^T R \hat{U} \quad (4.17)
\end{aligned}$$

Omitting the terms that are not dependent on the optimization variable $\hat{U}$ gives:

$$\begin{aligned}
& \hat{X}(k)^T \Phi^T Q \Gamma_u \hat{U} + \hat{U}^T \Gamma_u^T Q \Phi \hat{X}(k) + \hat{U}^T \Gamma_u^T Q \Gamma_u \hat{U} + \hat{U}^T \Gamma_u^T Q \Gamma_w \hat{W} + \hat{W}^T \Gamma_w^T Q \Gamma_u \hat{U} + \hat{U}^T R \hat{U} \\
&= \hat{U}^T \underbrace{\left(\Gamma_u^T Q \Gamma_u + R\right)}_{H} \hat{U} + 2 \underbrace{\left(\hat{X}(k)^T \Phi^T Q \Gamma_u + \hat{W}^T \Gamma_w^T Q \Gamma_u\right)}_{f} \hat{U} \quad (4.18)
\end{aligned}$$

Thus, the objective function can be rewritten as:

$$J = \min_{\hat{U}} \hat{U}^T H \hat{U} + 2f \hat{U} \quad (4.19)$$

Now the problem is formulated in such a way that a QP solver can be used, where the Hessian matrix (H) needs to be positive definite, f is the gradient vector. The typical optimization problem to solve in an MPC problem is formulated as

$$\begin{aligned}
J &= \min_{\hat{U}} \frac{1}{2} \hat{U}^T H \hat{U} + f \hat{U} \\
\text{s.t.} \quad & A \hat{U} \leq b \\
& lb \leq \hat{U} \leq ub \quad (4.20)
\end{aligned}$$

where the constraints can be formulated as a linear combination of the optimization variables, and also by upper and lower bounds on the optimization variables. The MATALB function quadprog is one of the ready-to-use solvers for solving quadratic objective functions with linear constraints.

4.3 Model Predictive Control (MPC)

To ensure the dissipativity constraint of semi-active suspension systems, the following constraint must be satisfied:

$$0 \leq c_{min} \leq c_{adaptive} \leq c_{max} \qquad (4.21)$$

The above dissipativity conditions can be transformed into input constraints.

$$\begin{cases} c_{min}(\dot{z}_{ui} - \dot{z}_{si}) \leq F_i \leq c_{max}(\dot{z}_{ui} - \dot{z}_{si}), & \text{if } (\dot{z}_{ui} - \dot{z}_{si}) \geq 0 \\ c_{max}(\dot{z}_{ui} - \dot{z}_{si}) \leq F_i \leq c_{min}(\dot{z}_{ui} - \dot{z}_{si}), & \text{if } (\dot{z}_{ui} - \dot{z}_{si}) < 0 \end{cases} \qquad (4.22)$$

Then, the dissipativity constraint is expressed as a set of linear inequalities between control input and state variables. Let the suspension deflection velocity $\Delta v_i = \dot{z}_{ui} - \dot{z}_{ci}$, the dissipativity constraint over the control horizon can be written as:

$$\text{If } \Delta v_i(k) \geq 0, \quad 0 \leq \underbrace{\begin{bmatrix} c_{min} \Delta v_i(k) \\ c_{min} \Delta v_i(k+1) \\ c_{min} \Delta v_i(k+2) \\ \vdots \\ c_{min} \Delta v_i(k+N_c-1) \end{bmatrix}}_{lb} \leq \underbrace{\begin{bmatrix} F_i(k) \\ F_i(k+1) \\ F_i(k+2) \\ \vdots \\ F_i(k+N_c-1) \end{bmatrix}}_{\hat{U}} \leq \underbrace{\begin{bmatrix} c_{max} \Delta v_i(k) \\ c_{max} \Delta v_i(k+1) \\ c_{max} \Delta v_i(k+2) \\ \vdots \\ c_{max} \Delta v_i(k+N_c-1) \end{bmatrix}}_{ub} \qquad (4.23)$$

$$\text{If } \Delta v_i(k) < 0, \quad \underbrace{\begin{bmatrix} c_{max} \Delta v_i(k) \\ c_{max} \Delta v_i(k+1) \\ c_{max} \Delta v_i(k+2) \\ \vdots \\ c_{max} \Delta v_i(k+N_c-1) \end{bmatrix}}_{lb} \leq \underbrace{\begin{bmatrix} F_i(k) \\ F_i(k+1) \\ F_i(k+2) \\ \vdots \\ F_i(k+N_c-1) \end{bmatrix}}_{\hat{U}} \leq \underbrace{\begin{bmatrix} c_{min} \Delta v_i(k) \\ c_{min} \Delta v_i(k+1) \\ c_{min} \Delta v_i(k+2) \\ \vdots \\ c_{min} \Delta v_i(k+N_c-1) \end{bmatrix}}_{ub} \leq 0 \qquad (4.24)$$

It can be seen that the constraints on the damping forces exerted by the semi-active dampers are dependent on the directions of suspension deflection velocities that are parts of the system states, which is hard to solve in real-time. There are a few solutions introduced in the literature to handle the constraint problems. In [20], three different approaches are introduced. First, relative damper velocities were predicted as a function of the initial values, road profile over the prediction horizon, and the damping force which is the optimization variable. In this expression, the relative velocities were calculated as outputs. Using the pre-measured, nonlinear damper characteristic that was represented by a polygonal line, the constraint on the damping force was obtained. This resulted in a linear quadratic optimization problem with nonlinear constraints. The optimization was solved at each time step by using the MATLAB function *fmincon*. Second, an appropriate linear damper force was chosen and then the relative damper velocities could be approximated. Constraints for the damping force were calculated over the prediction horizon which led to a linear optimization problem. In the third approach, optimal control signals were calculated without considering input constraints. If the obtained damping force was outside the feasible region, the force would be clipped and the maximum or minimum force would be applied. In this way, the resulting forces were no longer the optimal one as desired. In [22], based on a nonlinear half-car model, the control signal was written as a function of the state variables using a set membership approach, which was obtained offline (could not be calculated online due to the long computational time). Then the only online calculation was a simple evaluation of the approximated off-line calculated function. However, a model with only four states was used in this study. The necessary number of offline calculations would be much more difficult to cover with a more complex model. Another approach in [43] was to apply the road classification strategies together with the MPC algorithm. They used a Lidar to measure the road profile in front of the vehicle and the Fast Fourier Transform (FFT) was applied to analyze the frequency content of the road. Then, the road profiles were divided into different classes such as potholes, bumps, random roads, etc. The MPC optimization problem was solved offline for all road classes and the results were stored in lookup tables. During real-time operation, the control signals were recalled from pre-defined lookup tables based on the current state values and the road class. This method was validated by using a half-car model, which showed that the MPC developed in [43] provided better ride comfort than Skyhook and H_∞ control. There is another popular approach which has been employed several times in the literature. The system was modelled as a Mixed Logical Dynamical (MLD) hybrid model by using Hybrid Systems DEscription Language (HYSDEL) included in Multi-Parametric Toolbox (MPT). The non-linear and non-convex force constraint was approximated by a piecewise affine approximated with three segments. The results were obtained with the aid of software packages MATLAB and *Yalmip*. *HYSDEL* was developed by Torrisi and Bemporad to obtain *MLD* models from a high-level textual description of the hybrid dynamics. *HYSDEL* models are used in the hybrid toolbox of MATLAB for modeling, simulating, and verifying the safety properties of hybrid systems. Besides, it is also used for designing MPC controllers for linear systems with constraints and hybrid systems and determining equivalent piecewise affine control functions that can be immediately prototyped on hardware. By using the *MPT*, one can design, analyze, and deploy optimal controllers for constrained linear, nonlinear and hybrid systems. *Yalmip* is a modeling language for advanced description and solution of convex and non-convex optimization problems, allowing the user to concentrate on the high-level model. Based on this method, the authors of [23] developed an explicit MPC for a semi-active suspension system. While in [44], the authors considered a linear parameter varying (LPV) MPC approach. Through simulation, the online computation of the MPC-based controller proposed in [19] was examined, with the use of MATLAB, *Yalmip* toolbox, and *Gurobi* solver. The study [21] proposed MPC hybrid control tools as a way to obtain control laws with different degrees of optimality. Moreover, the authors of [24] introduced a new approach to deal with the semi-active dissipative constraint. In their work, a full-car model with the damping coefficient as the control signal was proposed. This resulted in a nonlinear model since the control signal was multiplied by some of the states. The benefit of formulating in this way was that the constraint on damping coefficients was

4.3 Model Predictive Control (MPC)

linear (only had a lower and an upper bound), but this approach resulted in a nonlinear MPC (NMPC) problem. The *ACADO* Toolkit was employed to handle this NMPC problem, which is an open-source framework for MPC and NMPC problems that utilizes *qpOASES* as its main low-level solver [29]. The proposed approach had been validated against a realistic vehicle model in *CarMaker*, which was able to run in real-time and the ride quality had been significantly improved compared to a passive suspension system.

An approach that is similar to one of the solutions mentioned in [20] can be used to easily deal with the semi-active damping dissipative constraint. For each time step, the previous optimal control signal $U^*(k-1)$ can be collected and then used for further system states. In order to linearize the dissipativity constraint problem, it can be assumed that the damping forces over the prediction and control horizons are the same. Under this assumption, further vehicle states can be obtained, which are then used for calculating the boundaries of the optimal control signals $U^*(k)$ for the next time step. Based on the above description, the proposed MPC control algorithm is concluded as follows:

MPC algorithm

Input: state vector $X(k)$, disturbance input vector $W(k)$, control signal vector $U^*(k-1)$

Output: optimal control signal $U^*(k)$

Assume that $U(k|k) = U(k+1|k) = \cdots = U(k+N_c-1|k) = U^*(k-1)$,
and $W(k|k) = W(k+1|k) = \cdots = W(k+N_c-1|k) = W(k-1)$.
Obtain $\hat{X}(k+1|k), \cdots, \hat{X}(k+N_p|k)$ according to Eq. 4.13.
Calculate the boundaries of the control signals according to Eqs. 4.23 and 4.24.
Obtain optimal solution $U^*(k)$ according to Eq. 4.20.
return the optimal control signal $U^*(k)$ for the next iteration.

Example 4.5 Develop an MPC algorithm from scratch in MATLAB for a semi-active suspension system.

Step 1: Design the system model based on the simplified full-car model, using the parameters of the mid-sized SUV given in Example 4.1.

Step 2: Develop the MPC algorithm as introduced in Eq. 4.9–4.24.

Step 3: Define the control output constraints. This example uses the off-the-shelf Continuous Damping Control (CDC) adaptive dampers produced by ZF. Its adjustable range is shown in Fig. 4.13. The damping characteristic can be adjusted by the current. Specifically, the lowest-damped and highest-damped states are achieved at 0A and 2A, respectively.

Step 4: Set up the co-simulation environment of MATLAB/Simulink and CarSim for high-fidelity simulation studies, as shown in Fig. 4.8. The offset-bump road profile used in this scenario is shown in Fig. 4.3, and the vehicle speed is 18 km/h.

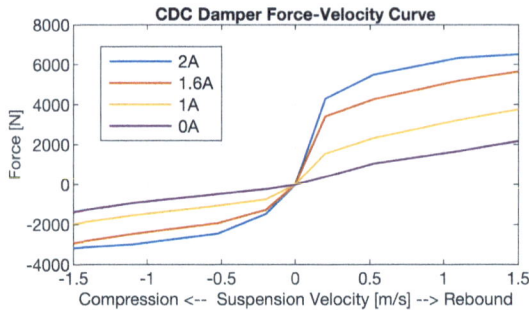

Fig. 4.13 CDC damper force–velocity curves

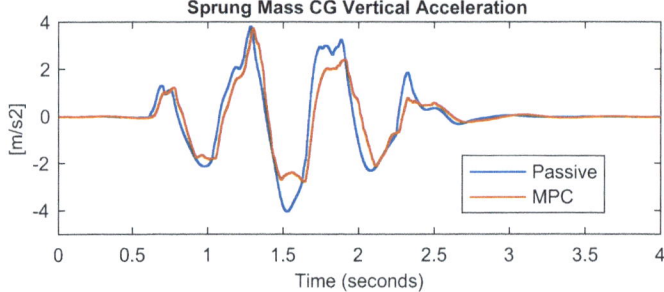

Fig. 4.14 Sprung mass vertical acceleration (Offset-bump at 18kph)

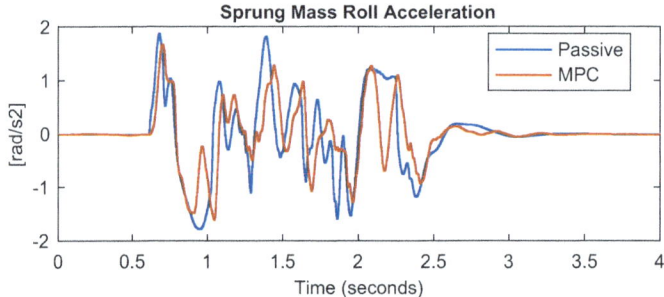

Fig. 4.15 Sprung mass roll acceleration (Offset-bump at 18kph)

The acceleration responses at the vehicle CG location in vertical, roll, and pitch directions are presented in Figs. 4.14, 4.15, and 4.16, respectively. In terms of the vertical and pitch weighted acceleration RMS values, MPC provides 27% and 22% reductions. In the roll direction, the MPC controller gives a 28% improvement. The damping forces of the passive and semi-active suspension systems are given in Figs. 4.17 and 4.18.

4.3 Model Predictive Control (MPC)

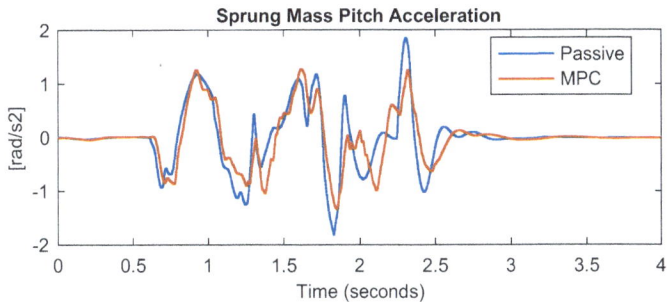

Fig. 4.16 Sprung mass pitch acceleration (Offset-bump at 18kph)

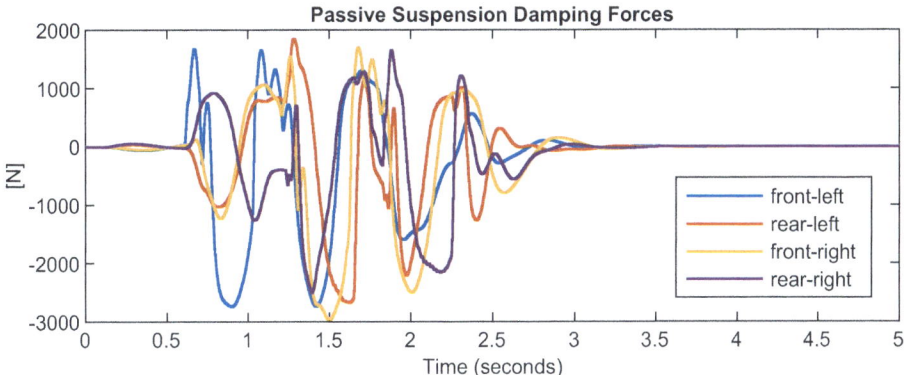

Fig. 4.17 Damping forces of the passive suspension

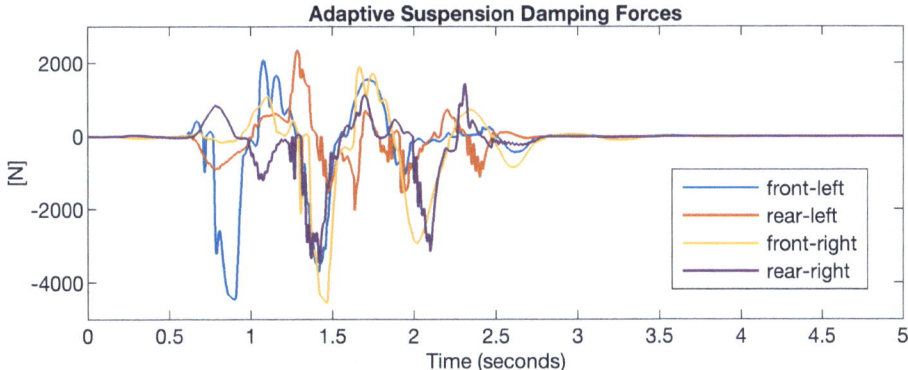

Fig. 4.18 Damping forces of the adaptive suspension using MPC

4.4 Integrated Skyhook-LQR

It is noticeable that MPC can properly handle the constrained multi-objective optimization problems, so it has great potential in vehicle suspension control tasks that are constrained by damping dissipative characteristics. However, its complicated algorithm requires a more powerful processor due to the high computational cost, which makes MPC expensive to implement on an automotive-grade microcontroller. Alternatively, the LQR is an option for realizing coordinated suspension control. It is a popular control technique used in both industry and academia to control linear dynamic systems because of its simplicity, optimality, integrability, and design flexibility. However, it cannot properly handle the damping constraints. Thus, developing a computationally efficient, robust, and effective suspension controller for practical applications becomes a key task in the automotive industry. As a solution, the Skyhook and LQR strategies are integrated to attenuate the undesired vibration in multiple directions [45]. As the name implies, the proposed Integrated Skyhook-LQR algorithm integrates the popular Skyhook approach to attenuate vertical vibration and the LQR approach to further optimize rotational dynamics. The reasons for choosing these two control approaches are listed as follows:

- Skyhook is the most widely used control method for vehicle suspension systems because of its simplicity, robustness, and efficiency. Using Skyhook as the base control algorithm can guarantee satisfactory ride quality and robustness. In addition, this approach considers the damping constraint in terms of the force applied direction, but not the force boundary.
- However, Skyhook cannot handle the multi-objective optimization problem, such as improving the vehicle dynamics in vertical, roll, and pitch directions at the same time. But LQR naturally seeks global optimum solutions for multi-objective problems. Thus, LQR is suitable for further optimizing the control performances.
- LQR requests less computational power than the MPC algorithm, so it is easier and cheaper for practical applications, especially the suspension control needs to be processed within the range of 100 Hz to 200 Hz.

The schematic of the proposed integrated Skyhook-LQR approach is shown in Fig. 4.19. In this algorithm, the control signal F_i is determined by both Skyhook and LQR modules. Based on the Skyhook control signal F_{sky}, the LQR module adjusts the final control signal F_i by the increment term ΔF_i. In this way, these two controllers work together to provide excellent coordinated control performances in vertical, pitch, and roll directions.

Example 4.6 Develop an Integrated Skyhook-LQR controller for a semi-active suspension system in MATLAB, using the parameters of the high-end sedan introduced in Example 4.2.

The damping characteristics of the passive suspension are assumed as when the current applies to the CDC damper is constantly 1A. The low-damped boundary is achieved when

4.4 Integrated Skyhook-LQR

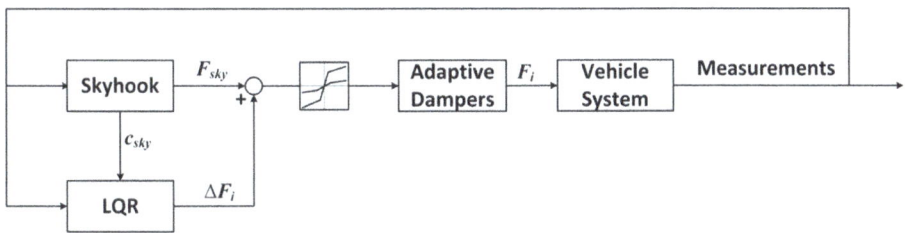

Fig. 4.19 Integrated Skyhook-LQR control

the current is 1.8A, and the high-damped boundary is achieved when the current is 0A. The force–velocity curves of the front CDC damper and rear CDC damper are shown in Figs. 4.20 and 4.21, respectively.

The output penalty matrix Q and the control signal penalty matrix R for the Skyhook-LQR controller are constant values regardless of the driving conditions. The following gains were used in this example:

$$Q = diag(0, 0, 0, 0, 100, 100)$$

$$R = 10^{-3} diag(1, 1, 1, 1)$$

By choosing different gain values for matrix Q, the control performance in vertical, roll, and pitch can be adjusted to a large extent. Hence, a proper weighting matrix can compromise the motions among all three directions to achieve an optimum solution for any

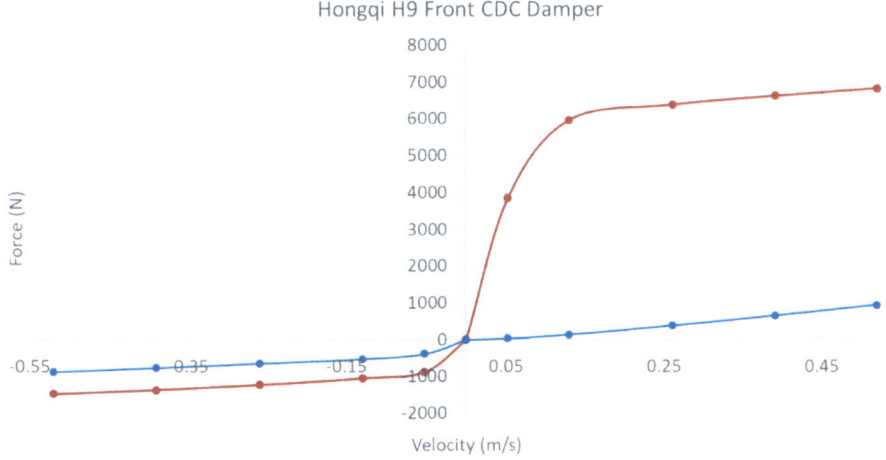

Fig. 4.20 Front CDC damper characteristics

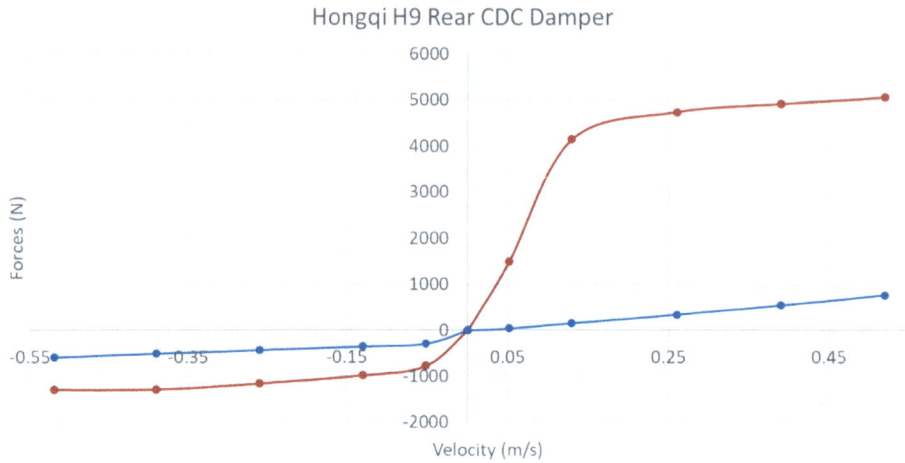

Fig. 4.21 Rear CDC damper characteristics

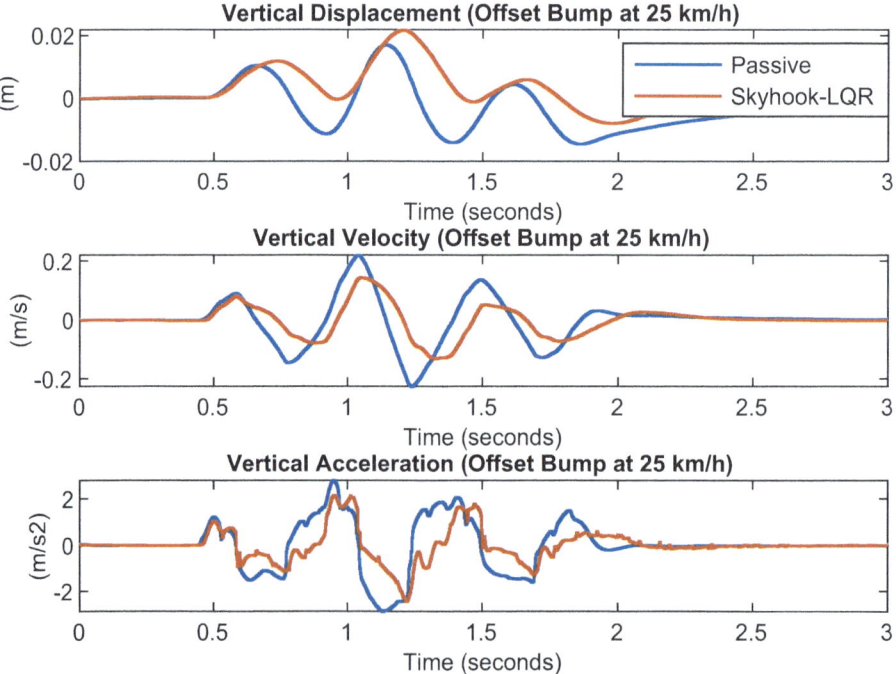

Fig. 4.22 Vehicle vertical motions at CG location

4.4 Integrated Skyhook-LQR

driving condition. The above gain values of the matrices Q and R were tuned by trial and error using the CarSim-Simulink co-simulations. The coefficient of the Skyhook algorithm was chosen as $c_{sky} = 5000\,Ns/m$. The offset-bump road profile used in this scenario is shown in Fig. 4.3, and the vehicle speed is 25 km/h.

The acceleration responses at the vehicle CG location in vertical, roll, and pitch directions are presented in Figs. 4.22, 4.23, and 4.24, respectively. In the vertical direction, the integrated Skyhook-LQR controller reduces the undesired vibrations by 23% in terms of the acceleration's RMS value. Its performance in roll direction is also significant, providing a 28% improvement. Similarly, this integrated approach reduces the pitch acceleration by 20%.

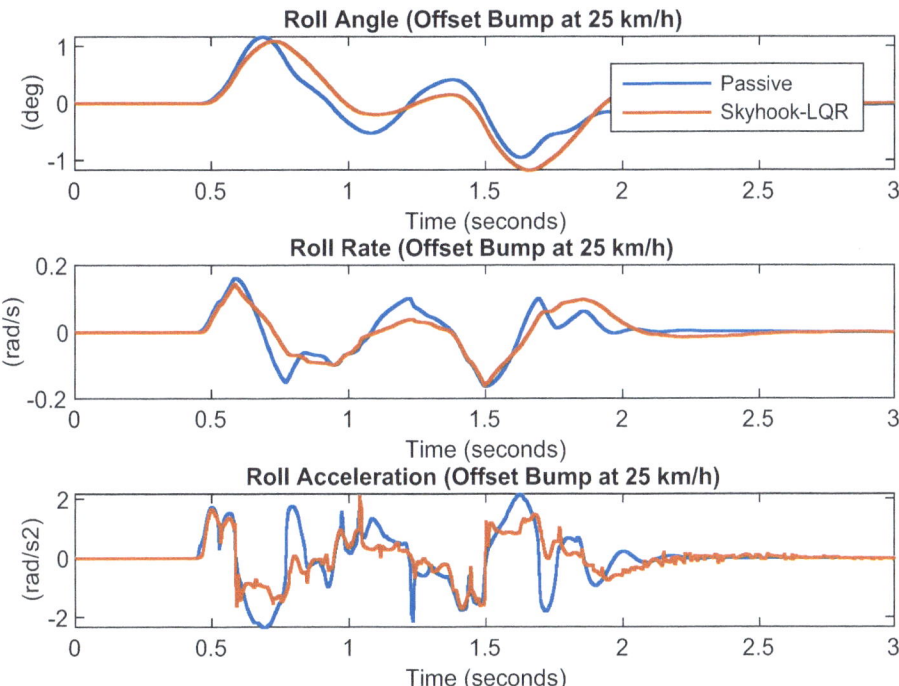

Fig. 4.23 Vehicle body motions in roll direction

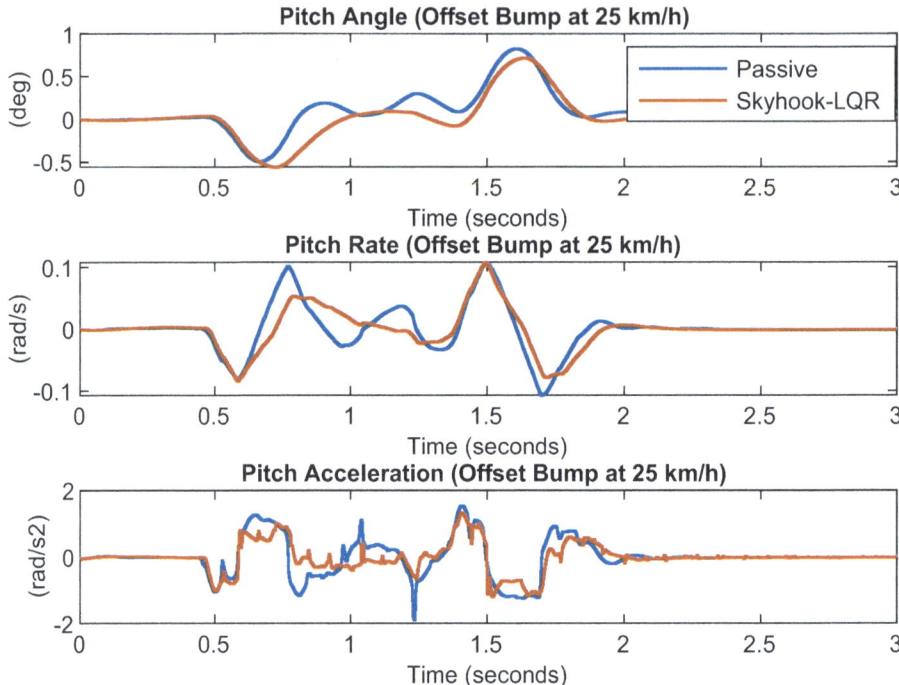

Fig. 4.24 Vehicle body motions in pitch direction

4.5 Gain-Adaptive Algorithms

By choosing different gain values for the output penalty matrix, the control objectives can be easily adjusted. Hence, a proper weighting matrix can compromise the motions among vertical, pitch, and roll directions to provide a better overall performance. However, it is not possible to use only one set of gains to achieve the optimum performance on various road conditions. For example, when tuning a suspension controller, higher weights are needed to attenuate significant disturbances, while lower weights are preferred when driving in better road conditions. This conflict makes the gain-adaptive technique necessary to continuously isolate the undesired vibrations as effectively as possible. Technically, one of the most commonly used solutions is to intelligently adjust the weighting matrices online based on the available sensor measurements. With the updated output penalty matrix, the gain-adaptive control algorithms can attenuate the vibrations in the human-sensitive range to the greatest extent.

Example 4.7 As shown in Fig. 4.25, assuming that there are two 6-axis IMUs are installed on the truck chassis frame and the cabin floor to detect their motions in real-time, respectively.

4.5 Gain-Adaptive Algorithms

Thus, the sensor measurements can be processed and analyzed to adjust the control gains. Two gain-adaptive algorithms are introduced in this example. Specifically, the first algorithm is processed according to the motions of the chassis frame (disturbance-based), while the second algorithm focuses on the dynamics of the cabin (state-based).

- **Disturbance-based gain-adaptive algorithm**

The structure of this disturbance-based gain-adaptive algorithm is presented in Fig. 4.26, which aims to select the optimal gains according to the result of its vibration level grading algorithm. These can be tuned by trial and error. The gain-adaptive algorithm is utilized to adjust the LQR weights for rapidly changed disturbances.

Based on the real-time IMU measurements, the vibration levels of the chassis frame in the pitch $\bar{\theta}_f$ and roll $\bar{\varphi}_f$ directions are graded based on the angular rates and the angular accelerations as follows:

$$\bar{\theta}_f = 5\big[abs(\dot{\theta}_f) + rms(a_{\theta w})\big]$$
$$\bar{\varphi}_f = 10\big[abs(\dot{\varphi}_f) + rms(a_{\varphi w})\big] \qquad (4.25)$$

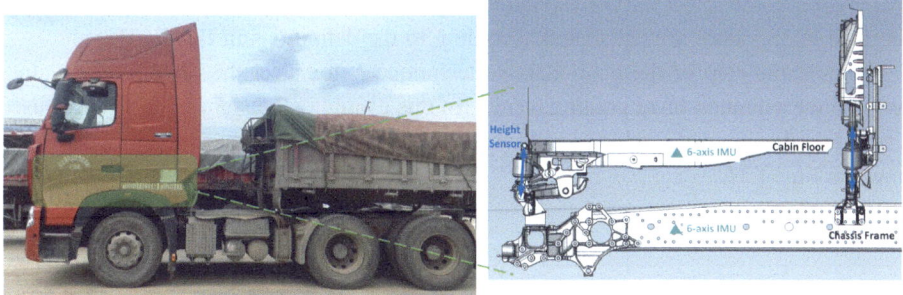

Fig. 4.25 Sensor locations

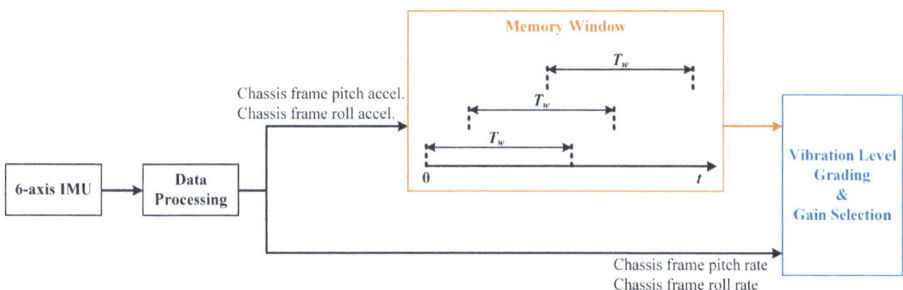

Fig. 4.26 Gain-adaptive algorithm

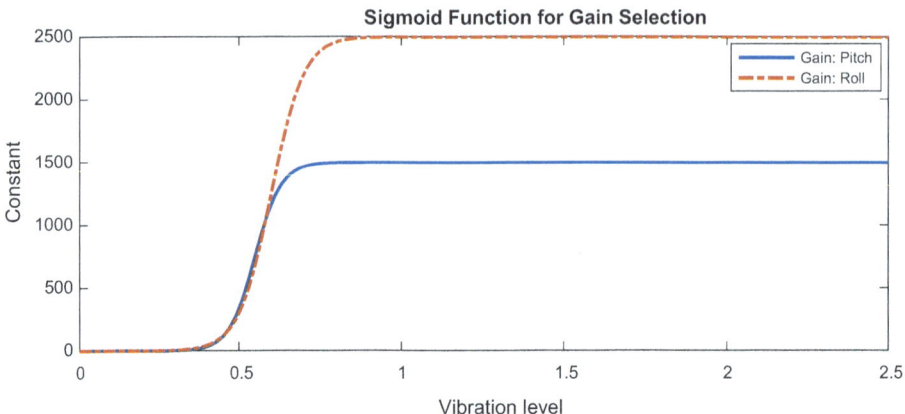

Fig. 4.27 The sigmoid function for gains selection

in which the absolute values of the pitch rate $\dot{\theta}_f$ and roll rate $\dot{\varphi}_f$ of the chassis frame are directly measured by a 6-axis IMU; the frequency-weighted root-mean-square (RMS) values of the pitch acceleration $a_{\theta w}$ and roll acceleration $a_{\varphi w}$ are calculated over a certain period of time T_w that is named the memory window. In detail, the measured angular rates are processed to get the corresponding angular accelerations, and then the frequency-weighted RMS values are calculated according to the definitions in ISO 2631-1.

The essential step of the gain-adaptive technique is to select the optimal gain values based on the vibration level grading results. In this example, the sigmoid function is used to schedule the gains of pitch and roll which guarantees a relatively smooth transition, as shown in Fig. 4.27.

$$q_{pitch} = \frac{1500}{1 + e^{-25*(\overline{\theta}_f - 0.55)}}$$
$$q_{roll} = \frac{2500}{1 + e^{-20*(\overline{\varphi}_f - 0.6)}} \qquad (4.26)$$

- **State-based gain-adaptive algorithm**

The state-based gain-adaptive algorithm selects the control gains based on the IMU measurement of the cabin motions, i.e., the cabin pitch rate $\dot{\theta}_s$ and its roll rate $\dot{\varphi}_s$.

As shown in Fig. 4.28, the trends of the cabin's pitch rate and roll rate are first defined as follows:

$$\dot{\theta}_{slope}(k) = sign\left[\dot{\theta}_s(k) - \dot{\theta}_s(k-1)\right]$$
$$\dot{\theta}_{dir}(k) = \dot{\theta}_{slope}(k) + \dot{\theta}_{slope}(k-1) \qquad (4.27)$$

4.5 Gain-Adaptive Algorithms

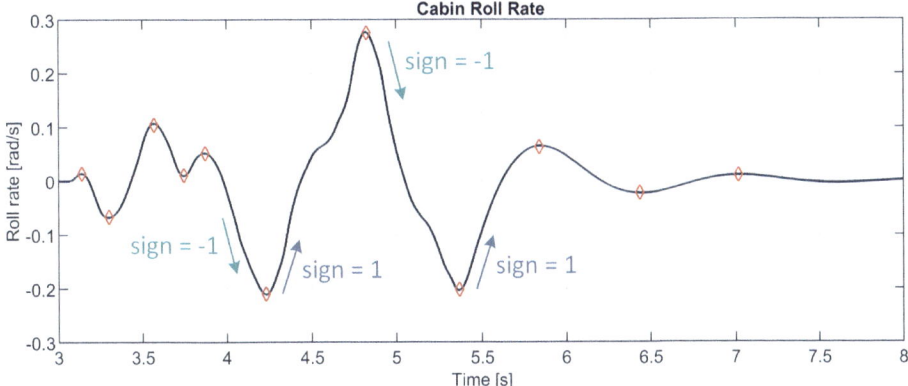

Fig. 4.28 Cabin roll rate response

$$\dot{\varphi}_{slope}(k) = sign[\dot{\varphi}_s(k) - \dot{\varphi}_s(k-1)]$$
$$\dot{\varphi}_{dir}(k) = \dot{\varphi}_{slope}(k) + \dot{\varphi}_{slope}(k-1) \quad (4.28)$$

If the sign of $\dot{\varphi}_{dir}(k)$ equals to 1 or -1, it means that the trend of the roll rate maintains the same. Otherwise, if $sign[\dot{\varphi}_{dir}(k)]$ equals 0, it means that the trend of the roll rate changes in this time step and reaches its local maximum value.

As a result, a coefficient is determined for pitch direction:

$$\dot{\theta}_{coef}(k) = abs[\dot{\theta}_s(k-1)], \text{ if } sign[\dot{\theta}_{dir}(k)] = 1 \text{ or } -1$$
$$\dot{\theta}_{coef}(k) = abs[\dot{\theta}_s(k)], \text{ if } sign[\dot{\theta}_{dir}(k)] = 0 \quad (4.29)$$

The same logic is used for the roll direction as follows:

$$\dot{\varphi}_{coef}(k) = abs[\dot{\varphi}_s(k-1)], \text{ if } sign[\dot{\varphi}_{dir}(k)] = 1 \text{ or } -1$$
$$\dot{\varphi}_{coef}(k) = abs[\dot{\varphi}_s(k)], \text{ if } sign[\dot{\varphi}_{dir}(k)] = 0 \quad (4.30)$$

Thus, the resulting control gain-adaptive algorithm for pitch $q_5(k)$ and roll $q_6(k)$ directions are

$$q_5(k) = \frac{1500}{1 + e^{-30[\dot{\theta}_{coef}(k) - 0.05]}}$$
$$q_6(k) = \frac{2500}{1 + e^{-30[\dot{\varphi}_{coef}(k) - 0.1]}} \quad (4.31)$$

The expected performances after implementing the proposed gain-adaptive algorithms on the adaptive suspension system are:

- When driving on bumpy roads, the gain-adaptive controller should be able to provide competitive performances as the non-adaptive controller. Since the gains of the non-adaptive controller are tuned for harsh road conditions.
- When driving on a paved road, the gain-adaptive controller should provide better ride quality than the non-adaptive controller, in which the vertical vibration is usually dominant.

It can be seen that the disturbance-based gain-adaptive algorithm calculates the frequency-weighted acceleration in real-time for grading the disturbance level, which spends more computational effort but fulfills the comfort criteria defined in ISO 2631. Differently, the state-based gain-adaptive algorithm utilizes the available state measurements, requiring less computational cost. In addition, the gain values generated by the disturbance-based gain-adaptive algorithm are more sensitive to the disturbance level, which sacrifices the system's robustness to some extent.

Diagnosis and Prognosis of Suspension Systems

Suspension operation reliability is a critical performance index in vehicle engineering, directly impacting maneuvering stability and driving safety [46]. Effective management of suspension systems is essential to ensure vehicle safety and performance. This chapter delves into the supervisory functions that monitor system integrity, detect undesirable conditions, and mitigate risks of damage or accidents. Supervisory tasks in suspension systems are vital for maintaining system health and ensuring safety. These tasks can be categorized into two main subjects:

- Fault detection and identification: fault detection involves continuous monitoring of critical suspension parameters to ensure they remain within acceptable limits. This process includes regular checks and diagnostic tests to identify any deviations from the norm that could indicate a potential fault.
- Fault diagnosis and prognosis: fault diagnosis aims to obtain the maximum amount of redundant information about the system's condition, and further filter and pinpoint the fault type with many details on root causes, and fault location. Fault prognosis focuses on forecasting system degradation and predicting potential failures and their timelines. By understanding the progression of wear and tear, maintenance teams can plan proactive interventions, thereby preventing unexpected breakdowns and extending the lifespan of suspension components.

The primary objective of these supervisory functions is to enable early detection of faulty behavior in suspension systems. By monitoring these systems in closed loops, any abnormal behavior can be promptly addressed, alerting the vehicle operator or maintenance teams to potential issues.

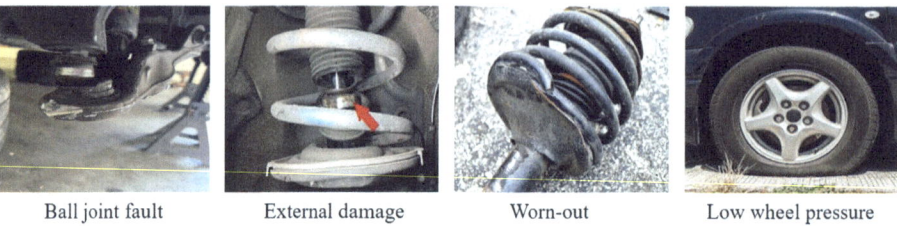

| Ball joint fault | External damage | Worn-out | Low wheel pressure |

Fig. 5.1 Common faults in the suspension structure

5.1 Common Component Failures in Suspension Systems

Figure 5.1 illustrates common component failures in a suspension system. Ball joint faults, often caused by driving on rough terrain, loss of lubrication, heavy loads, and frequent braking, manifest as wheel wobbling and steering wheel vibration. Strut external damage, typically resulting from physical impacts by foreign objects like stones, leads to dents and rigid suspension. Prolonged use, loss of lubrication, and rupture can cause a strut to worn-out, which may result in varying ride heights, a bouncy ride, steering pull to one side, and increased braking distance. Low wheel pressure results in stiff or hard steering and uneven tire wear.

Various factors such as wear, lack of lubrication, misalignment, heavy loads, mishandling, improper installation, and corrosion can exacerbate these faults. Early detection is crucial for maintaining suspension performance, minimizing maintenance disruptions, and preventing potentially dangerous accidents. Therefore, fault diagnosis is essential for ensuring the safety, reliability, and comfort of vehicle operation.

5.2 Suspension Fault Detection and Identification

Detecting if any abnormality happens and identifying the source of such fault is essential for many systems including suspension. In many studies, this step can be simplified as parameter estimation as the abnormality corresponds to the ill-posed system parameter, such as the spring and damper coefficients.

5.2.1 Fault Detection with Non-Recursive Parameter Estimation

Many control processes can be described by the regression model [47]:

$$y_i = b_0 + b_1 x_{i1} + b_2 x_{i2} \ldots + b_m x_{im} + \varepsilon_i, i = 1, 2, \ldots, n \tag{5.1}$$

5.2 Suspension Fault Detection and Identification

in which y_i is i-th response variable; x_i is the i-th predictor variable and b_i the estimated coefficient; and ε_i is the random error term.

The regression model can be written in the vector form as follows:

$$\begin{bmatrix} y_1 \\ y_2 \\ \vdots \\ y_n \end{bmatrix} = \begin{bmatrix} 1 & x_{11} & x_{12} & \cdots & x_{1m} \\ 1 & x_{21} & x_{22} & \cdots & x_{2m} \\ \vdots & \vdots & \vdots & \vdots & \vdots \\ 1 & x_{n1} & x_{n2} & \cdots & x_{nm} \end{bmatrix} \begin{bmatrix} b_0 \\ b_1 \\ \vdots \\ b_m \end{bmatrix} + \begin{bmatrix} \varepsilon_1 \\ \varepsilon_2 \\ \vdots \\ \varepsilon_n \end{bmatrix} \quad (5.2)$$

or simply

$$Y = X\beta + \varepsilon \quad (5.3)$$

The parameter vector β can be estimated by the non-recursive least square method. In the simple regression problem, the goal is to predict y_i from the knowledge of x_i. Given n pairs of data $[(x_1, y_1), \cdots, (x_n, y_n)]$ and the unknown parameters $[b_0, \cdots, b_m]$, one can obtain the Residual Sum of Squares (RSS) as

$$\text{RSS} = \sum_{i=1}^{n} (y_i - (b_0 + b_1 x_i + b_2 x_i + \ldots + b_m x_i))^2 \quad (5.4)$$

A perfect-fitting of the regression implies minimization of $y_i - (b_0 + b_1 x_i + b_2 x_i + \ldots + b_m x_i)$, hence the parameters can be obtained by solving the derivatives of RSS as 0. The general vector form of the solution can be obtained as

$$\hat{\beta} = \left(X^T X\right)^{-1} X^T Y \quad (5.5)$$

Example 5.1 Non-recursive algorithms can be applied to estimate parameters such as the damping coefficient (c), suspension stiffness (k), and other system characteristics from dynamic data. These parameters are essential for developing accurate models of the suspension system dynamics, which are represented by ordinary differential equations (ODEs).

Consider the ordinary differential equation for a sprung mass in a suspension system:

$$m_u \ddot{z}_u = -k(z_u - z_s) - c(\dot{z}_u - \dot{z}_s) - F_C \quad (5.6)$$

where m_u is the unsprung mass, z_s is the displacement of the body (sprung mass), z_u is the displacement of the wheel (unsprung mass), k is the stiffness coefficient, c is the damping coefficient and F_C is the coulomb friction.

To estimate the parameters k and c using the non-recursive algorithm, one can first collect measurements of z_u, $\dot{z}_u$, z_s, $\dot{z}_s$ over time from sensors. Assuming the F_C can be measured or estimated separately, one can construct the vector form where

$$Y = m_u \ddot{z}_u$$
$$X = \begin{bmatrix} z_s - z_u \\ \dot{z}_s - \dot{z}_u \end{bmatrix}$$
$$\beta = \begin{bmatrix} k \\ c \end{bmatrix} \tag{5.7}$$

For example, consider a nominal mass of 1000 kg, with stiffness of 15,000 N/m, and damping 1000 Ns/m, assuming there is no external force, the response of initial condition w_B at 0.1 m is plotted in the top left of Fig. 5.2. Similarly, the response with 10% parameter variation on mass, damping coefficient and bias are presented in the top right of Fig. 5.2.

With clean data collected during the nominal response, the estimated parameters exactly match the true values. However, if data is collected during an experiment where parameters suddenly change at 5 s, the estimated parameters often result in unreliable predictions (as shown in Fig. 5.3). This is because the data corresponds to two different systems: the nominal system before the change and the altered system after the change.

The non-recursive least squares algorithm assumes that the signals are stored during the measurement period and that the parameters are calculated all at once afterward. This non-recursive approach is particularly suitable for offline identification, where data can be analyzed post-experiment to identify system parameters.

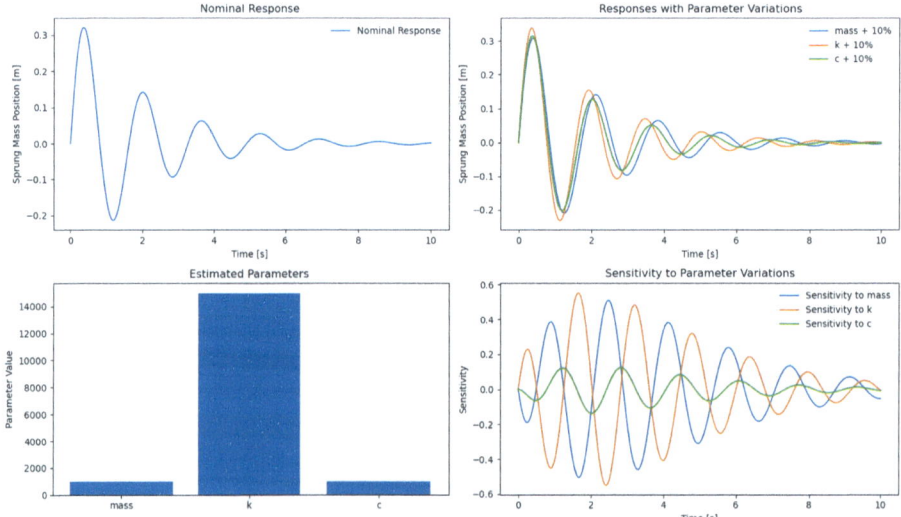

Fig. 5.2 System response, parameters, and the sensitivity

5.2 Suspension Fault Detection and Identification

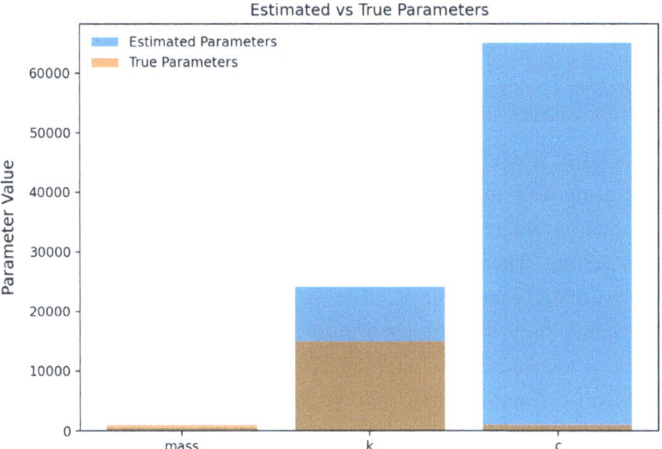

Fig. 5.3 Estimated parameters and the real parameters

Fault detection can be done by using non-recursive parameter estimation in suspension systems. When system parameters change abruptly, as in real-world scenarios, the non-recursive approach may struggle to provide accurate estimations due to the mixed data from different system states. Additionally, parameter sensitivity plays a significant role in the accuracy of fault detection. Variations in parameters such as mass, damping coefficient, and spring stiffness can significantly impact the system's dynamic response, highlighting the importance of continuous and adaptive parameter estimation methods for reliable fault detection and system diagnostics.

5.2.2 Fault Detection with Recursive Parameter Estimation

In practical applications, it is common to retrieve new data and the system parameter could vary during the operation. Using non-recursive least squares algorithms in such scenarios can be time-consuming. The Recursive Least Squares (RLS) algorithm with a forgetting factor is a popular method for parameter estimations, as described by [48]. This algorithm operates on the principle that older data is less reliable than newer data. Interested readers are encouraged to refer to relevant literature and MATLAB packages for detailed implementations [49, 50].

The Kalman filter is another powerful tool widely used for recursively estimating the internal states of a linear dynamic system from a series of noisy measurements. When applied to system parameter estimation, it adopts a linear regression form similar to the previous sections with $y(k) = \varphi^T(k)\theta(k) + e(k)$. In this case, the system parameters θ are no longer time-invariant, the parameter covariance update equation in the Kalman filter algorithm can be written as:

$$P(k+1) = P(k)\left[I - \frac{\varphi(k+1)\varphi^\top(k+1)P(k)}{1+\varphi^\top(k+1)P(k)\varphi(k+1)}\right] + R \quad (5.8)$$

where the process noise R is chosen as a positive definite matrix with each element along the diagonal indicating how much the parameter varies in a random walk model [51].

Consider a scenario where a vehicle experiences an unexpected additional load or partial suspension failure. The Kalman filter, set up for parameter estimation, would detect deviations in the estimated mass compared to its nominal value. If these deviations exceed a predefined threshold, they can trigger a fault alarm in the vehicle's diagnostic system. Figure 5.4 illustrates a case where a sudden change in the mass occurs, and the Kalman filter-based parameter estimator effectively tracks this change with a rapid response.

In MATLAB/Simulink, one can use the *Recursive Polynomial Model Estimator* block under the *System Identification Toolbox/Estimators* library, as shown in Fig. 5.5, to specify the model structure, initial estimate, and related options in the algorithm block. The Kalman filter can be considered as a special case with a forgetting factor equal to one [52].

Recursive parameter estimation methods, such as the RLS algorithm and the Kalman filter, offer robust solutions for real-time fault detection in suspension systems. These methods accommodate continuous data inflow and adapt to parameter variations dynamically, making them suitable for practical applications where system parameters may

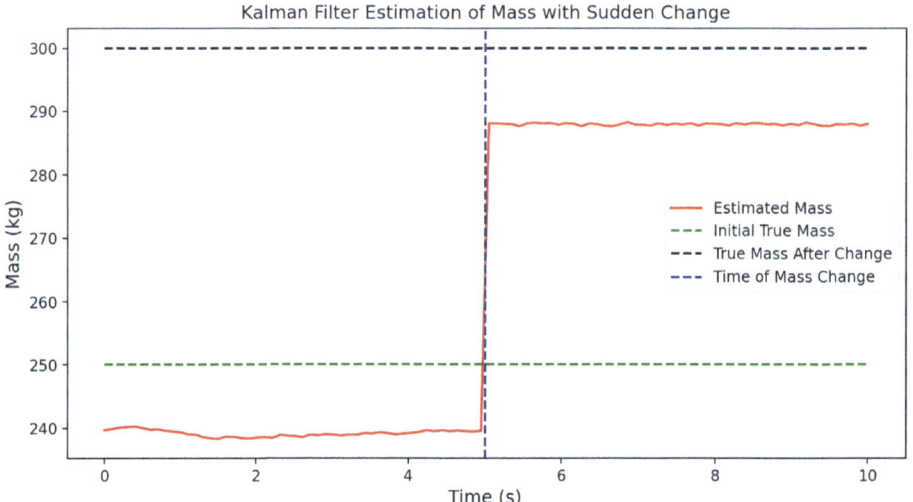

Fig. 5.4 Estimation with Kalman filter on sudden parameter change

5.3 Suspension Fault Diagnosis and Prognosis

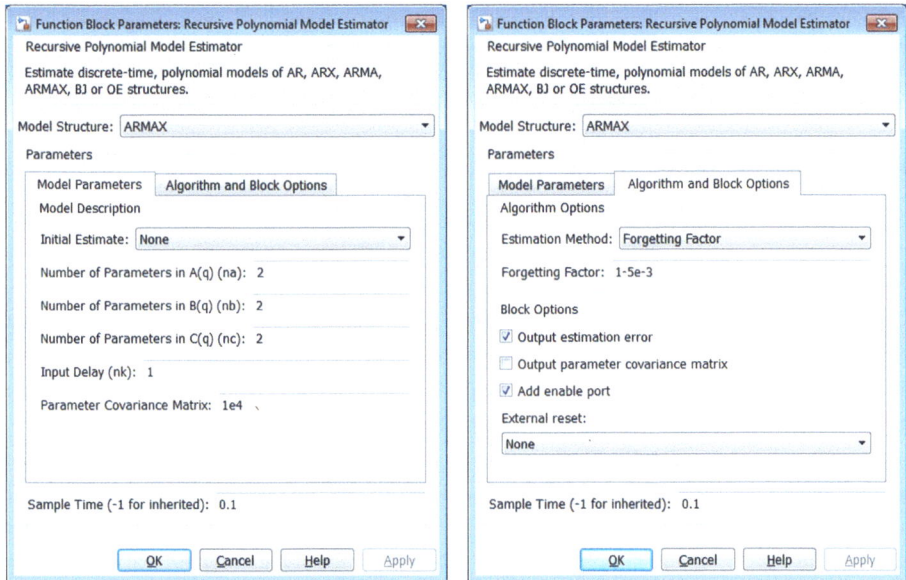

Fig. 5.5 MATLAB Simulink Recursive Polynomial Model Estimator block

change unexpectedly. By effectively tracking deviations in parameters like mass or damping coefficients, these algorithms enable timely fault detection, enhancing the safety and reliability of vehicle operations. The ability to process new data efficiently and respond to parameter changes promptly underscores the importance of recursive methods in modern diagnostic systems for automotive suspensions.

5.3 Suspension Fault Diagnosis and Prognosis

As the fault patterns are detected, it is often necessary to diagnose such faults, for example identifying the causes. Given enough data and performance degradation measures, potential fault prognosis can be helpful in improving suspension safety with proper condition examination and remaining life estimation.

5.3.1 Fault Diagnosis Procedure

The primary goal of any fault diagnosis approach is to obtain maximum redundant information about the system's condition, ideally without adding extra sensors. The diagnostic procedure relies on observed analytical and heuristic symptoms, along with heuristic knowledge of the process. In a fault-free scenario, it is preferred that diagnostic residuals

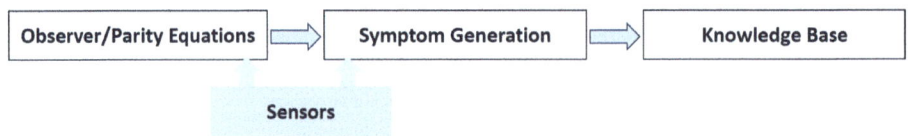

Fig. 5.6 Typical procedure for fault diagnosis

are zero, indicating no faults. However, due to disturbances and modeling uncertainties, residuals may deviate even when no faults are present.

The typical procedure is presented in Fig. 5.6, where the residual signal is generated by comparing the actual measure of the system and the estimated ones either using a model or inverse model. The non-zero mean residuals consist of the symptoms and then they are diagnosed by the expert or a knowledge base for further detailed analysis.

An example of an electronically controlled air suspension (ECAS) can be seen in Fig. 5.7, where the actuator is the air spring solenoid valve for each wheel corner. Through the failure analysis of the potential failure modes, the fault source and their effects can be drawn in Fig. 5.7. The simple example shows the air spring solenoid valve fault tree for each spring (front-left, front-right, rear-left, rear-right) that includes short/open circuit, spool block, and potential spring fatigue. According to [53], the failure mode analysis is carried out for each component and the threshold for each signal pattern is determined for the fault detection step. The common adaptive thresholds for the failure modes are decided based on linearization error and parameter uncertainty error [54].

Take the example in Table 5.1, the residue R2 is completely insensitive to fault modes 1 and 2. With the significant level of R2, we have high confidence to diagnose the system in fault mode 3. On the other hand, R1 is sensitive to both fault modes 1 and 2. Similarly, for R3, the fault diagnosis and isolation step is expected to be performed by matching and decoupling the fault columns to the residuals.

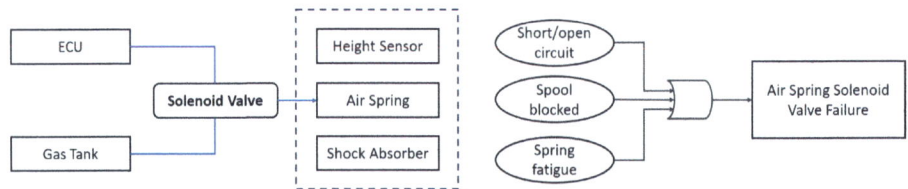

Fig. 5.7 A Demonstration of electronically controlled air suspension system and its fault tree

Table 5.1 Example of influence structure between fault mode and residuals

	Fault mode 1	Fault mode 2	Fault mode 3
R1	1	1	0
R2	0	0	1
R3	0	1	1

5.3.2 Outlooks on Fault Prognosis

Fault prognosis is crucial as it provides the expected occurrence of a fault in the future. Along with the fault diagnosis process, the fault prognosis is also aimed at improving the system's performance in the long run. One recent trend in the prognosis problem for suspension systems is the prediction of remaining useful life (RUL) that can provide helpful decision support for maintenance [55, 56]. While directly predicting an expected time to the system's end-of-life could be challenging, monitoring the system degradation would be the one important feature to reveal the evolutionary trend caused by the various fault sources.

Suspension fault prognosis is a critical aspect of vehicle maintenance, directly impacting safety and performance. While current methods like residual analysis and recursive estimation provide a solid foundation, integrating advanced technologies and adaptive algorithms will drive future advancements. For example, developing an adaptive algorithm that can adjust to changing operating conditions for fault detection, diagnosis and prognosis in real-time will effectively improve the safety monitoring process for the vehicle chassis domain. Implementing predictive maintenance based on the prognosis can minimize the costs and prevent unexpected failures. Continuous research and development in this field will lead to more accurate, reliable, and efficient fault detection and prognosis, ultimately enhancing vehicle safety and longevity.

References

1. F.D. Goncalves, M. Ahmadian, A hybrid control policy for semi-active vehicle suspensions. Shock Vibrat. **10**(1), 59–69 (2003)
2. C. Liu et al., Generalized Skyhook-Groundhook hybrid strategy and control on vehicle suspension. IEEE Trans. Vehicular Technol. (2022)
3. S.M. Savaresi, E. Silani, S. Bittanti, Acceleration-Driven-Damper (ADD): An optimal control algorithm for comfort-oriented semiactive suspensions (2005)
4. S. Nie et al., A semi-active suspension control algorithm for vehicle comprehensive vertical dynamics performance. Veh. Syst. Dyn. **55**(8), 1099–1122 (2017)
5. J. Theunissen et al., Preview-based techniques for vehicle suspension control: a state-of-the-art review. Annual Rev. Control **51**, 206–235 (2021)
6. H. Li, X. Jing, H.R. Karimi, Output-feedback-based H∞ control for vehicle suspension systems with control delay. IEEE Trans. Ind. Electron. **61**(1), 436–446 (2013)
7. H. Du, N. Zhang, L. Wang, Switched control of vehicle suspension based on motion-mode detection. Veh. Syst. Dyn. **52**(1), 142–165 (2014)
8. P. Li, J. Lam, K.C. Cheung, Control of vehicle suspension using an adaptive inerter. Proc. Inst. Mech. Eng. Pt. D: J. Automobile Eng. **229**(14), 1934–1943 (2015)
9. X. Zheng et al., Active full-vehicle suspension control via cloud-aided adaptive Backstepping approach. IEEE Trans. Cybern. **50**(7), 3113–3124 (2020).
10. X. Tang et al., Takagi–Sugeno fuzzy control for semi-active vehicle suspension with a magnetorheological damper and experimental validation. IEEE/ASME Trans. Mechatron. **22**(1), 291–300 (2016)
11. M. Zhang, X. Jing, G. Wang, Bioinspired nonlinear dynamics-based adaptive neural network control for vehicle suspension systems with uncertain/unknown dynamics and input delay. IEEE Trans. Ind. Electron. **68**(12), 12646–12656 (2020)
12. S.D. Nguyen, B.D. Lam, S. Choi, Smart dampers-based vibration control–Part 2: fractional-order sliding control for vehicle suspension system. Mech. Syst. Signal Process. **148**, 107145 (2021)
13. S.D. Nguyen, S. Choi, J. Kim, Smart dampers-based vibration control–Part 1: measurement data processing. Mech. Syst. Signal Process. **145**, 106958 (2020)
14. S. Lu et al., Integrated control on MR vehicle suspension system associated with braking and steering control. Veh. Syst. Dyn. **49**(1–2), 361–380 (2011)
15. M. Čorić et al., Optimisation of active suspension control inputs for improved vehicle ride performance. Veh. Syst. Dyn. **54**(7), 1004–1030 (2016)
16. J. Na et al., Active suspension control of quarter-car system with experimental validation. IEEE Trans. Syst. Man Cybern. Syst. (2021)

17. G. Koch, T. Kloiber, Driving state adaptive control of an active vehicle suspension system. IEEE Trans. Control Syst. Technol. **22**(1), 44–57 (2014). https://doi.org/10.1109/TCST.2013.2240455
18. E. Pellegrini, *Model-Based Damper Control for Semi-Active Suspension Systems*. Technical University of Munich (2012)
19. M.Q. Nguyen et al., A Model Predictive Control approach for semi-active suspension control problem of a full car. *Cdc,* pp. 721–726 (2016)
20. C. Gohrle et al., Model Predictive Control of semi-active and active suspension systems with available road preview. *Ecc,* pp. 1499–1504 (2013). https://doi.org/10.23919/ECC.2013.6669185
21. N. Giorgetti et al., Hybrid model predictive control application towards optimal semi-active suspension. Int. J. Control **79**(5), 521–533 (2006). https://doi.org/10.1080/00207170600593901
22. M. Canale, M. Milanese, C. Novara, Semi-active suspension control using "fast" model-predictive techniques. Tcst **14**(6), 1034–1046 (2006). https://doi.org/10.1109/TCST.2006.880196
23. L.H. Cseko, M. Kvasnica, B. Lantos, Explicit MPC-based RBF neural network controller design with discrete-time actual Kalman filter for semiactive suspension. Tcst **23**(5), 1736–1753 (2015). https://doi.org/10.1109/TCST.2014.2382571
24. K. Fredrik, S. Simon, *Real-Time Nonlinear Model Predictive Control for Semi-Active Suspension with Road Preview*. Chalmers University of Technology (2018)
25. J. Theunissen et al., Regionless explicit model predictive control of active suspension systems with preview. Tie **67**(6), 4877–4888 (2020). https://doi.org/10.1109/TIE.2019.2926056
26. M.M. Morato et al., Design of a fast real-time LPV model predictive control system for semi-active suspension control of a full vehicle. J. Franklin Inst. **356**(3), 1196–1224 (2019). https://doi.org/10.1016/j.jfranklin.2018.11.016
27. J. Wu et al., Ride comfort optimization via speed planning and preview semi-active suspension control for autonomous vehicles on uneven roads. IEEE Trans. Vehicular Technol. **69**(8), 8343–8355 (2020)
28. C. Gohrle et al., Active suspension controller using MPC based on a full-car model with preview information. *Acc,* pp. 497–502 (2012). https://doi.org/10.1109/ACC.2012.6314680
29. H.J. Ferreau et al., qpOASES: a parametric active-set algorithm for quadratic programming. Math. Prog. Comp **6**(4), 327–363 (2014). https://doi.org/10.1007/s12532-014-0071-1
30. G. Cimini, A. Bemporad, D. Bernardini, ODYS QP Solver (2017)
31. K. Cheshmi et al., NASOQ: numerically accurate sparsity-oriented QP solver. ACM Trans. Graphics **39**(4), 96:1–96:17 (2020). https://doi.org/10.1145/3386569.3392486
32. B. Stellato et al., OSQP: an operator splitting solver for quadratic programs. Math. Prog. Comp **12**(4), 637–672 (2020). https://doi.org/10.1007/s12532-020-00179-2
33. J. Cao et al., State of the art in vehicle active suspension adaptive control systems based on intelligent methodologies. Tits **9**(3), 392–405 (2008). https://doi.org/10.1109/TITS.2008.928244
34. L.H. Nguyen, K. Hong, S. Park, Road-frequency adaptive control for semi-active suspension systems. Int. J. Control Autom. Syst. **8**(5), 1029–1038 (2010). https://doi.org/10.1007/s12555-010-0512-1
35. R. Kosut, Suboptimal control of linear time-invariant systems subject to control structure constraints. Tac **15**(5), 557–563 (1970). https://doi.org/10.1109/TAC.1970.1099555
36. K. Hong, H. Sohn, J.K. Hedrick, Modified skyhook control of semi-active suspensions: a new model, gain scheduling, and hardware-in-the-loop tuning. J. Dyn. Sys. Meas. Control **124**(1), 158–167 (2002). https://doi.org/10.1115/1.1434265
37. K. Yi, B.S. Song, A new adaptive sky-hook control of vehicle semi-active suspensions. Proceedings of the institution of mechanical engineers. Part D, J. Automobile Eng. **213**(3), 293–303 (1999). https://doi.org/10.1243/0954407991526874

38. G. Koch, K.J. Diepold, B. Lohmann, *Multi-objective Road Adaptive Control of an Active Suspension System*. Anonymous Dordrecht: Springer Netherlands, pp. 189–200 (2008)
39. I. Fialho, G.J. Balas, Road adaptive active suspension design using linear parameter-varying gain-scheduling. Tcst **10**(1), 43–54 (2002). https://doi.org/10.1109/87.974337
40. D. Karnopp, M.J. Crosby, R.A. Harwood, Vibration control using semi-active force generators. J. Eng. Indus. **96**(2), 619–626 (1974). https://doi.org/10.1115/1.3438373
41. S. Di Cairano, I.V. Kolmanovsky, Real-time optimization and model predictive control for aerospace and automotive applications, *Acc,* pp. 2392–2409 (2018). https://doi.org/10.23919/ACC.2018.8431585
42. Y. Zhang, *Multi-axle Vehicle Modeling and Stability Control: A Reconfigurable Approach* (2019)
43. M. Ahmed, F. Svaricek, Preview control of semi-active suspension based on a half-car model using Fast Fourier Transform. *Ssd,* pp. 1–6 (2013)
44. M.M. Morato, O. Sename, L. Dugard, LPV-MPC fault tolerant control of automotive suspension dampers. IFAC-PapersOnLine **51**(26), 31–36 (2018). Available: https://www.sciencedirect.com/science/article/pii/S2405896318328350. https://doi.org/10.1016/j.ifacol.2018.11.172
45. Y. Lu et al., A new integrated Skyhook-LQR coordinated control approach for semi-active vehicle suspension systems. J. Dyn. Sys. Meas. Control, pp. 1–10, (2022). https://doi.org/10.1115/1.4056442
46. C. Sun et al., Toward ensuring safety for autonomous driving perception: standardization progress, research advances, and perspectives. IEEE Trans. Intell. Transp. Syst. (2023)
47. S.C. Rutan, Recursive parameter estimation. J. Chemometrics **4**(2), 103–121 (1990)
48. T. Söderström, L. Ljung, I. Gustavsson, A theoretical analysis of recursive identification methods. Automatica **14**(3), 231–244 (1978)
49. Recursive Least Squares Estimator. Available: https://www.mathworks.com/help/ident/ref/recursiveleastsquaresestimator.html
50. R. Majjad, Estimation of suspension parameters. In: *Proceedings of the 1997 IEEE International Conference on Control Applications* (1997)
51. K.B. Singh, S. Taheri, Integrated state and parameter estimation for vehicle dynamics control. Int. J. Vehicle Perform. **5**(4), 329–376 (2019)
52. Linear Model Identification. Available: https://www.mathworks.com/help/ident/linear-model-identification.html?s_tid=CRUX_lftnav
53. X. Jiang, X. Xu, H. Shan, Model-based fault diagnosis of actuators in electronically controlled air suspension system. World Electric Vehicle J. **13**(11), 219 (2022)
54. Z. Zhang, X. He, Active fault diagnosis for linear systems: within a signal processing framework. IEEE Trans. Instrumen. Measur. **71**, 1–9 (2022)
55. A. Abou Jaoude, Analytic and linear prognostic model for a vehicle suspension system subject to fatigue. Syst. Sci. Control Eng. **3**(1), 81–98 (2015)
56. J. Luo et al., Model-based prognostic techniques applied to a suspension system. IEEE Trans. Syst. Man Cybern Part A: Syst. Humans **38**(5), 1156–1168 (2008)

SPRINGER NATURE

GPSR Compliance

The European Union's (EU) General Product Safety Regulation (GPSR) is a set of rules that requires consumer products to be safe and our obligations to ensure this.

If you have any concerns about our products, you can contact us on ProductSafety@springernature.com

In case Publisher is established outside the EU, the EU authorized representative is:

Springer Nature Customer Service Center GmbH
Europaplatz 3
69115 Heidelberg, Germany

The manufacturer's authorised representative in the EU is Springer Nature Customer Service Centre GmbH, Europaplatz 3, 69115 Heidelberg, Germany. If you have any concerns regarding our products, please contact ProductSafety@springernature.com

Printed and bound by CPI Group (UK) Ltd, Croydon, CR0 4YY

23/03/2026

02076360-0020